Minitab Manual
Michael G. Sklar

SIXTH EDITION

BASIC BUSINESS STATISTICS
Concepts and Applications

Mark L. Berenson

David M. Levine

Prentice Hall, Englewood Cliffs, New Jersey 07632

Production Editor: *Naomi Sysak*
Supplement Acquisitions Editor: *Diane Peirano*
Manufacturing Buyer: *Ken Clinton*

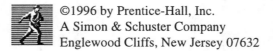©1996 by Prentice-Hall, Inc.
A Simon & Schuster Company
Englewood Cliffs, New Jersey 07632

Printed in the United States of America

10   9   8   7   6   5   4   3   2

ISBN 0-13-359506-4

Prentice-Hall International (UK) Limited, *London*
Prentice-Hall of Australia Pty. Limited, *Sydney*
Prentice-Hall Canada Inc., *Toronto*
Prentice-Hall Hispanoamericana, S.A., *Mexico*
Prentice-Hall of India Private Limited, *New Delhi*
Prentice-Hall of Japan, Inc., *Tokyo*
Simon & Schuster Asia Pte. Ltd., *Singapore*
Editora Prentice-Hall do Brasil, Ltda., *Rio de Janeiro*

Index of Problems
by Chapter

Index of Problems

Index of Problems

INTRODUCTION TO MINITAB

(There are no exercises from Chapter 1 of the textbook.)

This manual is primarily directed toward *Minitab for Windows*, which provides a menu system to access commands, rather than the *session command* approach of typing commands directly at the *Minitab* prompt MTB>. However, most of the instructions in this manual provide procedures for both versions. Therefore this manual may be profitably used for those who employ the *Windows* version of *Minitab*, as well as for those who do not use the *Windows* version.

The Minitab Worksheet

The *Minitab for Windows* worksheet provides several windows, which may be moved and sized to satisfy individual preference. One approach for setting up the display is shown in the graphic on the next page. This is a screen capture from the *Minitab* worksheet, depicting in one possible configuration the four different windows which are set up.

1. The *Session Window*: the upper left window is where commands are typed (when the menus are not used), and where the results of commands are displayed. Note the information box in this window, which describes how the window is used. (This box doesn't appear on the screen; it was added for information purposes.)

2. The *Data Window*: the lower left window is where the columns of data reside. Data may be edited in this window. Note the descriptive box explaining the purpose of this window.

    a. Also note the column headings: *Prices, Subscrpt, American,* and *NY*. When the instructions in this manual suggest typing a heading for a column, you may type the heading directly into the cell below the column designation (the column designations which Minitab uses are C1, or C2, and so on). However, you may also type a command into the Session Window at the Minitab prompt, which instructs *Minitab* to enter the heading; this approach will be described later.

3. The *History Window:* As the descriptive box indicates, the *History Window* stores the various commands which have been executed, either through the menus or through the Session Window. These commands may be copied from the *History Window* (using the menu choices *Edit, Copy* or using keyboard equivalents such as *Ctrl-C*) and then pasted into the *Session Window* to be executed there.

4. The *Info Window:* This window provides at-a-glance information about the data which are in the *Data Window.* Values of computed variables such as K1 are displayed here.

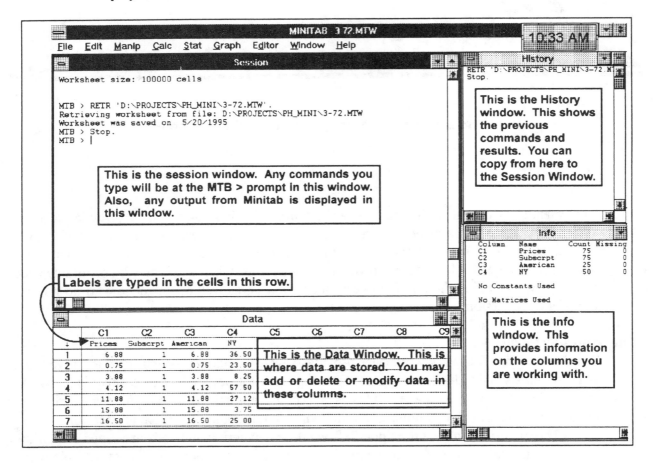

MENUS

The *Minitab for Windows* screen contains a menu structure similar to other windows applications. For instance, the *File* menu contains menu items which let you read in a Minitab worksheet, read in an ASCII file, save a file, print, and so on. As the graphic above shows, the main Minitab menu choices are: File, Edit, Manip, Calc, Stat, Graph, Editor, Window, and Help. These menu items (and their subitems) are accessible by clicking with the mouse, or by holding the Alt key and pressing the character which is underlined. (For instance, to open the *File* menu, you could hold the Alt key and press the letter "f" on the keyboard, which is the underlined character in the menu item File.)

1.   The *File* menu contains submenu choices as shown in the graphic below. Some of the submenu choices are described below.

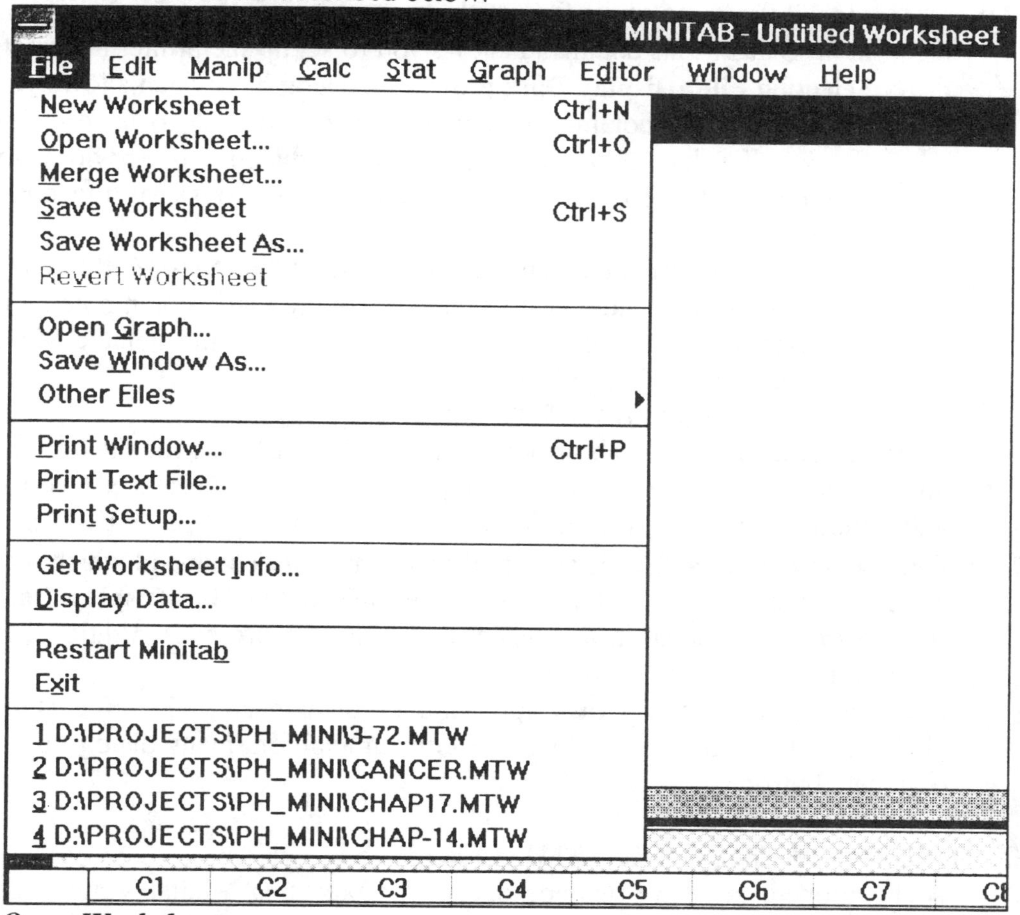

	MINITAB - Untitled Worksheet
File **Edit** **Manip** **Calc** **Stat** **Graph** **Editor** **Window** **Help**	

```
New Worksheet                           Ctrl+N
Open Worksheet...                       Ctrl+O
Merge Worksheet...
Save Worksheet                          Ctrl+S
Save Worksheet As...
Revert Worksheet

Open Graph...
Save Window As...
Other Files                                  ▶

Print Window...                         Ctrl+P
Print Text File...
Print Setup...

Get Worksheet Info...
Display Data...

Restart Minitab
Exit

1 D:\PROJECTS\PH_MINI\3-72.MTW
2 D:\PROJECTS\PH_MINI\CANCER.MTW
3 D:\PROJECTS\PH_MINI\CHAP17.MTW
4 D:\PROJECTS\PH_MINI\CHAP-14.MTW
```

C1	C2	C3	C4	C5	C6	C7	C8

2.   ***Open Worksheet***
To open a worksheet, use the menu **File>Open Worksheet** (that is, click on *File*, then click on *Open Worksheet)*, which provides a dialog box for you to choose which worksheet. Note that this menu selection pertains to a Minitab worksheet, with a file extension of *MTW*, such as *DATA.MTW*. A Minitab worksheet is stored in a particular format which is not readable by other programs.

3.   ***Save Worksheet***
Saving a worksheet in the Minitab format requires using the menu selection **File>Save Worksheet**. (The *Save Worksheet As* menu item provides a dialog box to enable you to save a file under a different name.)

4.   ***ASCII Files***
Many of the data files used in this manual are ASCII files; files with no formatting. In order to use these files, select **File>Other Files>Import ASCII Data** and provide the name and location of the ASCII file.

5.   ***Printing and Displaying***
There are three choices for printing, and several approaches for displaying data:

a. *Print Window*: This prints the contents of whichever window is active; Session, Data, Info, Graph, or History.

b. *Print Text File*: This command provides a dialog box which allows you to specify the name of the file which you wish to print.

c. *Print Setup*: This command allows you to set up the printer information.

d. Selecting **File>Display Data** provides a dialog box which allows you to specify the name of the column or columns which you wish to have displayed in the *Session Window*. For instance, when the dialog box provides you a list of columns from which to select, you may wish to select the column labeled *Prices*.

    1. The equivalent command for typing at the *Session Window* (instead of using the menus) is shown below. Note that a column name must have a leading and trailing apostrophe (except for the default Minitab column names C1, C2, and so on).
```
MTB >print 'Prices'
```

6. *Labeling columns*: Minitab provides its own default labels for the columns; C1, C2, and so on. When we use our own, more descriptive, labels, we must use a leading and trailing apostrophe, such as in the previous command, `print 'Prices'`

7. *Get Worksheet Info:* This is an extension of the *Print* command which may print data to the Session Window. The output is the same as the **File>Display Data** command (or the *Print* command if you typed it), but it also provides additional information, such as the count of the column.

8. *Help:* as with most windows applications, an extensive Help file is available, by clicking on the main menu selection *Help*. In addition, many dialog boxes have their own help buttons.

USING MENUS VERSUS TYPING COMMANDS

1. As mentioned, there are two approaches for executing Minitab commands: the menu approach and the direct-typing approach.

    a. Suppose you wish to obtain some summary information (mean, median, and so on) about column C1, which has the label *Hours*.

b.　The ***menu*** approach is

Select **Stat>Basic Statistics>Descriptive Statistics**

which means to first click on the main menu choice *Stat*, then when a second tier of menus is presented in a list, select or click on *Basic Statistics*, and when a third tier of menus is presented in a list, select or click on *Descriptive Statistics*.

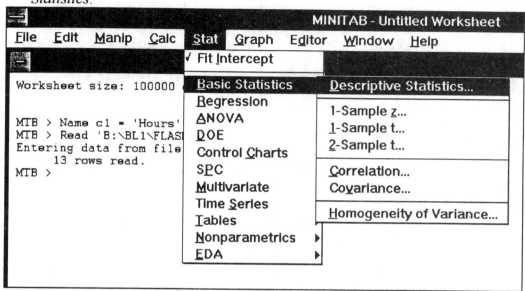

c.   At this point a dialog box will appear, an example of which is shown below. The instructions would ask you to double-click on the column name *Hours* in the box to the left (the box on the left has both the Minitab designation C1 and the label we've given it, *Hours*, which together constitute one choice*)*, and Minitab would then duplicate that name in the box under the heading *Variables*. (In the graphic below, we've already double-clicked on the item in the left box, **C1   Hours**   , and Minitab has already copied the heading to the box under *Variables*.   Once you click *OK*, the output would appear in the *Session Window*.)

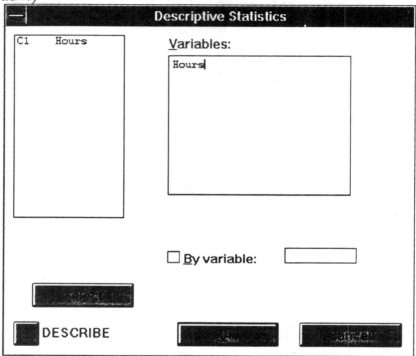

2.   The output appears in the *Session Window*, as shown in the graphic on the next page. As you may notice, although we used the menus to generate the *Describe* command, Minitab also placed the command in the *Session Window*. In addition, the output of the *Describe* command is shown. In other words:
a.   The menu items were selected: **Stat>Basic Statistics>Descriptive Statistics**, and the column label *Hours* was selected, and *OK* was clicked.
b.   Minitab typed for us in the *Session Window* the same command we would have typed ourselves if we had not used the menus: **Describe 'Hours'.**

c.   Minitab provided in the *Session Window* the output from the command, which begins with the heading **Descriptive Statistics.**

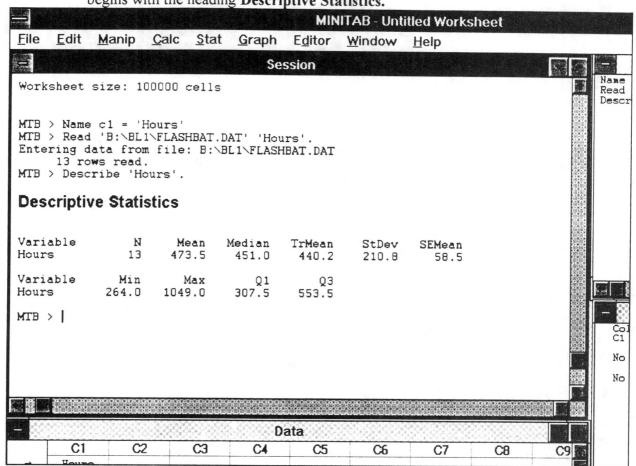

3.   **Typing:** If you had typed the command at the Minitab prompt in the *Session Window* rather than using the menus, you would have typed the bolded portion below (ignoring the Minitab prompt, which Minitab already placed in the Session Window):
MTB > **Describe 'Hours'.**

4.   The output from typing the *Describe* command would have been the same as you obtained from using the menu choices **Stat>Basic Statistics>Descriptive Statistics.**

TYPING DATA DIRECTLY INTO THE DATA WINDOW

	C1	C2	C3	C4	C5
↓					
1					
2					

Note the arrow (↓) pointing down. Clicking on the arrow will toggle it down or right.

1. Throughout this manual, the predominate way of entering data is to read in an ASCII file. However, data may be entered directly into the *Data Window*.

    a. A portion of the *Data Window* is shown in the graphic above. In the box to the left is an arrow pointing down (↓). This indicates that when the *Enter* key is pressed, the cursor automatically moves *down*, in the direction of the arrow. Therefore, you may type a number, press the *Enter* key, and the cursor will move *down* to the next cell.

    b. If, instead, you wanted the cursor to move to the *right* after entering a number (or an alphanumeric value) into a cell, then the arrow must be pointing to the right (→). The arrow may be toggled down or right by clicking on it in the box.

2. To enter data, simply click on the cell you wish to populate with a value and type the value and press *Enter*.

3. Of course, you may use the arrow keys on your keyboard to move around the cells in the *Data Window*, once you have moved to the *Data Window* by clicking somewhere within the *Data Window*.

MODIFYING DATA

1. To modify a value in a cell, click on that cell in the *Data Window* and simply type the new value, which will replace the previous value. Or, you can use the arrow keys on your keyboard to move to the cell to be edited and type to replace the existing entry.

COLUMN HEADINGS

1. Throughout this manual are suggestions to type descriptive headings for columns, since it is much easier to read the output with a descriptive label for the column rather than C1, for instance. Suppose we wish to label column C1 as *Hours*.

    a. **Typing label directly into the cell:** You may type the column label directly into the *Data Window*, below the Minitab designator such as C1, C2, and so on. You don't need apostrophes surrounding the label if you type it directly.

    b. **Typing label at the Session Window:** You may also define the name or label for a column by instructing Minitab to enter the name in the *Data Window*. At the Minitab prompt at the *Session Window*, use the command *name* to define the label for column C1 as *Hours*. Be sure to enclose the label name with leading and trailing apostrophes. At the Minitab prompt (which is already present; you don't type it), type what is in bold:

        ```
        MTB > name c1 'Hours'
        ```

INCLUDING COMMENTS WITH COMMANDS

On occasion you may wish to include a comment along with a command, to provide more information for someone else or to clarify a command for yourself.

a.   The symbol # lets Minitab know that what follows is to be treated as a comment and not part of a command.

b.   For example, using the previous command which provided the name *Hours* for column C1, if you wished to include a comment with this command you would simply append the # to the command, followed by your comment.

c.   `MTB > name c1 'Hours'  #Creates label Hours for C1`

GRAPH TICK MARKS

a.   Often when constructing a graph of some sort the user will be instructed to set *Tick Marks* for the graph. Tick marks are the small lines on the axes which provide an indication of where the data values occur. There are Major tick marks (the larger lines) and Minor tick marks (the smaller lines between the larger lines). The tick marks may be set up differently for each axis.

b.   To indicate the approach, the Cancer data used for various exercises will be described. Similar exercises are covered in more detail in Chapter 3.

c.   For the example problem, we assume the Cancer data are in column c1, and the goal is to provide a histogram. Select the menu choices **Graphs>Histogram**, and the cursor is in a cell in the *Graph Variables* section. Type **c1** (or the name of the column containing the Cancer data). Click on **Frame>Tick,** and a screen appears which is similar to that shown below. The bolded numbers enclosed in the boxes are the ones to enter. In Row 1, enter **6** for the Number of Major ticks and enter **4** for the number of Minor Ticks. In Row 2, enter **5** and **4**. (Initially the default values may be *Auto*: just delete these and enter the numbers as shown.) Then Click OK twice.

	Direction	Side	Positions	Number of Major	Number of Minor
1	X			6	4
2	Y			5	4

d.   The resulting graph is shown below.

1.   Notice that for the X axis we've specified 6 *Major* tick marks, and they are labeled 250, 300, 350, 400, 450, and 500. There are 4 *Minor* tick marks between each *Major* tick mark. Four Minor tick marks were selected to divide the distance between the Major tick marks into reasonable numbers. Here they represent units of 10, so the Major tick mark is at 250, then the Minor tick marks are at 260, 270, 280, 290, and the next Major tick mark is at 300. The goal is to be able to easily estimate the location of the data values.

2. For the Y axis there are 5 Major tick marks, at 0, 5, 10, 15, and 20. There are 4 Minor tick marks, each with a value of 1. The initial Major tick mark is at 0, and the minor tick marks are then 1, 2, 3, 4, and then the next Major tick mark is at 5. Again, this makes it easier to analyze the data being graphed.

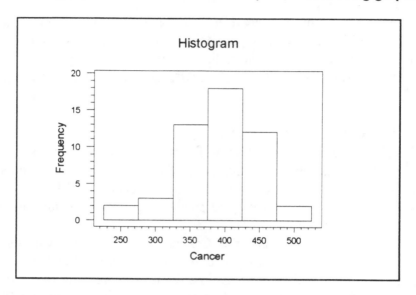

e. Now suppose instead of specifying the tick marks we let Minitab do it automatically. Instead of specifying Major tick marks and Minor tick marks, we would use the default *Auto* for each. Notice that Minitab used 2 Major and no Minor tick marks for the Y axis. Minitab specified 6 Major and no Minor tick marks for the X axis. For many graphs, the *Auto* default would work fine, but for most of the graphs in this workbook, we've specified the various tick marks for ease of analysis. (Note also that we've included a title for each graph; how to accomplish this will be discussed in a later section.)

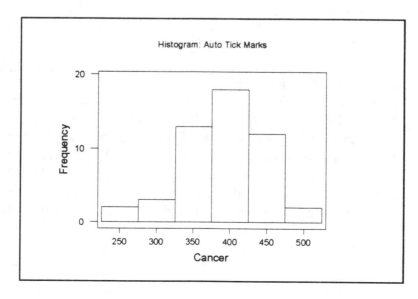

f.   (The tick marks shown in this section differ slightly from the ones shown in Chapter 3, even though the data are the same, and the tick mark dialog box previously described is filled out the same way. The difference is that in Chapter 3 we also introduced an *additional feature*: specifying the definition of the intervals and the associated midpoint locations. In that example we wanted to match the character graph we obtained from *Minitab*, so we specified particular midpoints.)

EDITING GRAPHS

1.   For many of the graphs, particularly the ones pertaining to the normal distribution in Chapters 8 through 11, additional features have been added. For instance, in the graph at right, a number of things have been added, including shading, arrows, labeling on the axes, and probability values. While a detailed discussion of this Minitab capability is beyond the scope of this manual, the general approach may be described. When a graph is generated and displayed in its own window, you may double-click the graph and the Minitab Graph Editor will be launched. Two tool palettes appear. (You may want to maximize the window.)

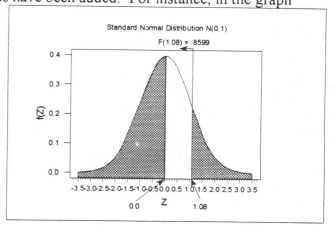

a.   To enter text, click on the "T" icon located on the smaller palette, move to where you wish the text to be, click and hold and move the mouse until the text box is the right size, and release. At the dialog box which appears type the text. You may then size and move the text box.

b.   To create a line and arrow, click on the "\" icon on the smaller palette, move to where you want the line to begin, click and hold and move until the line is where you want it and release. Click on the "arrows" icon on the larger palette and click on one of the arrows in the group of possible arrow configurations.

c.   Shading is a little trickier. Click on the icon on the smaller palette which looks like a crown. Move to the shape you want to shade. Click repeatedly on the outline of the shape (moving the mouse between clicks to a new spot on the outline). When you have come full circle (or whatever the shape is), the shape will close (it will be a closed shape). Click on the hatching icon on the larger palette and select the shading you desire.

d.   The main menu item *Editor* has a *View* subitem and an *Edit* subitem (and others which pertain to graph editing). When you have finished editing a graph, click on the *View* subitem.

DATA
COLLECTION

The importance of data collection, where data can be thought of as numerical information, is fundamental. This chapter describes methods and motivations for collecting data.

(There are no exercises from the textbook for this chapter.)

PRESENTING NUMERICAL DATA IN TABLES AND CHARTS

This chapter describes methods for organizing and effectively presenting, in the form of tables and charts, large batches of numerical data, with the goal of enhancing data analysis and interpretation.

PROBLEM 3.6
The data pertain to the incidence of cancer in all 50 states.

If the data aren't currently loaded, open the file by using the menu **File>Other Files>Import ASCII Data**; type **c1** for *column* and click *OK*. Give the file name *cancer.dat* and its location. Type the heading **Cancer** for column c1 and **Original** for column c2.

- a.   To demonstrate sorting: **Manip>Sort**; double-click on *original* in the left box so *original* appears in the *Sort Column(s):* box.
- b.   Click in the *Store sorted column(s) in:* box and double-click on *Cancer*.
- c.   Click in the first *Sort by column* box and either type **c2** there or again double-click *Original* in the left box.
- d.   If you typed rather than using the menus (the #comment is optional),
  ```
  MTB > Sort 'Original' 'Cancer'; #include the semicolon
  SUBC> By 'Original'.              #include the period
  ```

e.   The following is the frequency distribution (which is not displayed by Minitab)

Cancer per 100,000	Midpoint	Number of States in this range	Percentage of States in this range
less than 260	240	2	4
260 but less than 300	280	1	2
300 but less than 340	320	5	10
340 but less than 380	360	13	26
380 but less than 420	400	13	26
420 but less than 460	440	12	24
460 but less than 500	480	3	6
500 but less than 540	520	1	2
		50	100

PROBLEM 3.15

f.   For the same data, problem 3.15: The following provides a histogram, using a *character* graph, a simple graph which doesn't use high-resolution graphics and therefore is available on any computer. Generate the character graph by **Graphs>Character Graphs>Histogram**; double-click on *Cancer* or c1 and click on *OK*. The character histogram is displayed.

```
Character Histogram
Histogram of Cancer    N = 50
Midpoint    Count
     240        2   **
     280        1   *
     320        5   *****
     360       13   *************
     400       13   *************
     440       12   ************
     480        3   ***
     520        1   *
```

If you typed rather than using the menu,
```
MTB > GStd.
MTB > Histogram  'Cancer'.
```

g.   (Note that Minitab automatically, at the end of the command, types **Gpro** to restore the high-resolution graphics.)

h.   Now produce a histogram, using high-resolution output:
Select **Graphs>Histogram**, double-click on *Cancer* in the box to the left or type C1.
For Display, Item 1 should show Display *Bar* for each Graph.

i.   Title: Click on **Annotation>Title** and type **Histogram for Problem 3.15**

j.   Click on *Options*. For *Type of Histogram* select *Frequency;* For *Type of Intervals* select *Midpoint;* For *Definition of Intervals: Midpoint/Cutpoint Positions* type *240:520/40*, which indicates the first midpoint is at 240, the last is at 520, and the intervals are 40 wide. (This corresponds to the previous Character Graph.)

k.   Axis tick marks: Click on **Frame>Tick** and you will see a screen with the following headings. The numbers enclosed in a box are the ones to enter. In Row 1, enter **6** for the Number of Major ticks and enter **4** for the number of Minor Ticks. In Row 2, enter **4** and **4**. (Initially the default values may be *Auto*: just delete these and enter the numbers as shown.) Then Click OK.

	Direction	Side	Positions	Number of Major	Number of Minor
1	X			6	4
2	Y			4	4

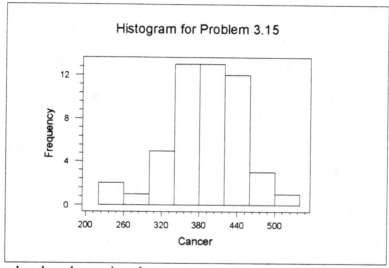

Histogram for Problem 3.15

l.   If you typed rather than using the menus,

```
MTB > Histogram 'Cancer';
SUBC>    MidPoint 240:520/40;
SUBC>    Bar;
SUBC>    Title "Histogram for Problem 3.15";
SUBC>    Axis 1;
SUBC>    Axis 2;
SUBC>    Tick 1;
SUBC>       Number 6 4;
SUBC>    Tick 2;
SUBC>       Number 4 4.
```

m.   Another character graph similar to the Histogram is the *DotPlot*, which typically is useful for smaller data sets. It will often have a dot for each observation, and a colon to represent two observations in one space.

    1.   Select **Graphs>Character Graphs>DotPlot**, double-click on *Cancer* in the box at left or type **c1**.

    2.   Ignore the other choices on the dialog box.

    3.   Click *OK*.

4.   The output is as shown below.
```
Character Dotplot
```

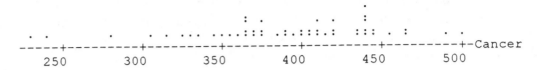

5.   If you typed rather than using the menus,
```
MTB > DotPlot 'Cancer'.
```

n.   Stem and Leaf: Select **Stat>EDA>Stem-and-Leaf**; double-click *Cancer* or *C1*. This displays the Stem & Leaf in the Session Window.

o.   The numbers to the left are the cumulative counts from the smallest number (229) at the top of the graph up to the class containing the median. A different sequence also ends at the class containing the median, but it starts at the bottom of the graph, going from the largest number (500).

p.   Shown on the next page are two *Stem-and-leaf* displays. There are many ways to display data using Stem-and-leaf: See the text for additional clarification.

1.   The display on the left has *leaf unit* of 10 (as stated on the display), so the original number 229 is represented as 2 2, where the left-most 2 stands for 200 and the right-most 2 is in units of 10, so it stands for 20. Thus the number 229 is split up into its stem of 2 (unit=100) and its leaf of 2 (unit=10) so the number represented is not the original 229 but is 220 (2 100's and 2 10's). Therefore some accuracy is lost by using the leaf unit of 10. Note also that *Minitab* has provided 4 categories for numbers in the 200's, and 5 categories for numbers in the 300's, and so on.

2.   The Stem-and-leaf on the right has *leaf unit* of 1. Here the original number 229 is accurately represented. In effect the number 22 9 means there are 9 1's, and 22 10's (which is 220), so 22 9 is the representation of 229. Note that when the units are 1's, the data are more spread out. The individual analyst must decide how to display the data and in what units.

q.   Following are the *Stem-and-Leaf* displays.

Character Stem-and-Leaf Display

Character Stem-and-Leaf Display

Stem-and-leaf of Cancer N = 50
Leaf Unit = 10

```
  2    2 23
  2    2
  2    2
  3    2 8
  5    3 01
  8    3 223
 11    3 445
 21    3 6666677777
 (5)   3 89999
 24    4 00000011
 16    4 223333
 10    4 444445
  4    4 66
  2    4 9
  1    5 0
```

Stem-and-leaf of Cancer   N = 50
Leaf Unit = 1.0

```
  1    22 9
  2    23 8
  2    24
  2    25
  2    26
  2    27
  3    28 2
  3    29
  4    30 7
  5    31 3
  7    32 69
  8    33 6
 10    34 58
 11    35 5
 16    36 04667
 21    37 12567
 22    38 3
 (4)   39 0168
 24    40 236889
 18    41 48
 16    42 22
 14    43 3488
 10    44 02235
  5    45 4
  4    46 34
  2    47
  2    48
  2    49 1
  1    50 0
```

r.   If you typed rather than using the menus,
```
MTB > Stem-and-Leaf 'Cancer';
SUBC>    Increment 1.
```

Problem 3.29
If the data aren't currently loaded, open the file by using the menu **File>Other Files>Import ASCII Data**; type **c1** for *column* and click *OK*. Give the file name *cancer.dat* and its location. Type the heading **Cancer** for column c1.

a.   Percentage Polygon
b.   Select **Graph>Histogram**
c.   For Graph Variables:, double-click *Cancer*

d.   In order for the Polygon to have both a symbol (circle) and for the symbols to be connected, the table must be filled out as shown below, where the **bold** is to be as shown.

For Display:

Item	Display	For each
1	**Symbol**	**Graph**
2	**Connect**	**Graph**

e.   Title: Select **Annotation>Title** and type **Percentage Polygon for Problem 3.29**

f.   Axis tick marks: Click on **Frame>Tick** and you will see a screen with the following headings. The numbers enclosed in a box are the ones to enter. In Row 1, enter **6** for the Number of Major ticks and enter **5** for the number of Minor Ticks. In Row 2, enter **4** and **4**. (Initially the default values may be *Auto*: just delete these and enter the numbers as shown.) Then Click OK.

	Direction	Side	Positions	Number of Major	Number of Minor
1	X			6	5
2	Y			4	4

g.   Click on the *Options* button. For *Type of Histogram* select *Percent;* For *Type of Intervals* select *Midpoint;* For *Definition of Intervals: Midpoint/Cutpoint Positions* type *240:520/40*, which indicates the first midpoint is at 240, the last is at 520, and the intervals are 40 wide.

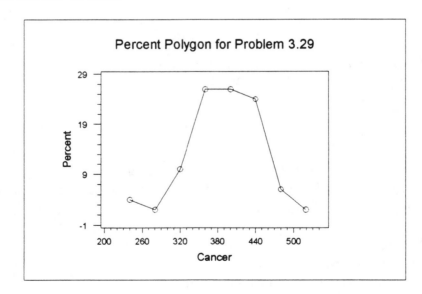

h. If you typed rather than using the menus,

```
MTB > Histogram 'Cancer';
SUBC>    Percent;
SUBC>    MidPoint 240:520/40;
SUBC>    Symbol;
SUBC>    Connect;
SUBC>    Title "Percentage Polygon for Problem 3.29";
SUBC>    Axis 1;
SUBC>    Axis 2;
SUBC>    Tick 1;
SUBC>       Number 6 5;
SUBC>    Tick 2;
SUBC>       Number 4 4.
```

Problem 3.37

Ogive (cumulative percentage polygon): This is the same *Cancer* data as shown previously.
   a. Follow the same instructions as for Problem 3.29 except:
      1. Click on the *Options* button and:
         a. For *Type of Histogram* click on *Cumulative Percent*
         b. For *Type of Intervals* select *Cutpoint;* For *Definition of Intervals: Midpoint/Cutpoint Positions* type *200:520/40*, which indicates the first interval begins at 200, the last ends at 520, and the intervals are 40 wide.
      2. Axis tick marks: Click on **Frame>Tick** and you will see a screen with the following headings. The numbers enclosed in a box are the ones to enter. In Row 1, enter **6** for the Number of Major ticks and enter **5** for the number of Minor Ticks. In Row 2, enter **4** and **5**. (Initially the default values may be *Auto*: just delete these and enter the numbers as shown.) Then Click OK.

	Direction	Side	Positions	Number of Major	Number of Minor
1	X			6	5
2	Y			4	5

3.   Modify the Title to *Cumulative Percentage Polygon (Ogive) for Problem 3.37*

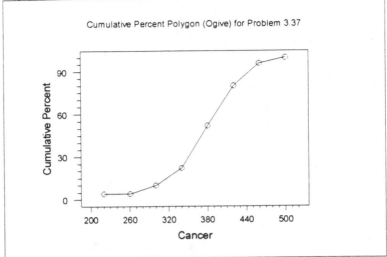

b.   If you typed rather than using the menus,

```
MTB > Histogram 'Cancer';
SUBC>    Cumulative;
SUBC>    Percent;
SUBC>    Cutpoint 200:540/40;
SUBC>    Symbol;
SUBC>    Connect;
SUBC>    Title "Cumulative Percentage Polygon (Ogive) for
    Problem 3.37";
SUBC>    Axis 1;
SUBC>    Axis 2;
SUBC>    Tick 1;
SUBC>      Number 6 5;
SUBC>    Tick 2;
SUBC>      Number 4 5.
```

PROBLEM 3.72:

To describe how to combine two graphs: these are closing stock prices for 25 issues on the American Exchange and 50 issues on the New York Exchange. The goal is to place the percentage polygon for each exchange on the same plot.

There are two excerises here: one produces a percentage polygon and the other produces a cumulative percentage polygon. The only difference in the two occurs in the title of the graph, and in the selection of a *percent* versus a *cumulative percent*.

In each case the user is instructed to overlay two graphs. It is advisable to turn this choice off after completing the two excerises.

If the data aren't currently loaded, open the file by using the menu **File>Other Files>Import ASCII Data**; type **c1-c2** for *column* and click *OK*. Give the file name *nyseax.dat* and its location. Type the heading **Prices** for column c1 and **Subscrpt** for column c2. The subscripts in column *Subscrpt* are 1 for *American* prices and 2 for *NY* prices.

Before constructing the two graphs we note that analyzing by frequencies may be misleading, since each sample is a different size; therefore we will use percentages for the analysis. The prices for American and the prices for NY are in the same column of the data file. Since the prices are stacked on top of each other (the first 25 are American and the next 50 are NY), they have to be split up so that American is considered as a group by itself, and so is NY. This requires using the subscripts in the second column labeled *Subscrpt*. This will be described below.

UNSTACKING
First, we need to *Unstack* the prices which are contained in column c1, by using the *Unstack* command.
   a.   Select **Manip>Unstack**
   b.   Click in the box below the heading *Unstack* and double-click on the *Prices* column in the box at the left.
   c.   Click in the box to the right of *Using subscripts in* and double-click on the *Subscrpt* column in the box at the far left.
   d.   Click in the first box below *Store results in blocks* and type **c5** so the American prices will be stored in column c5. Tab to the next box down and type **c6** so the NY prices will be stored in column c6. Click *OK*.
   e.   Type the label for column c5: **Price** and type the label for column c6: **NY.** (The heading for column c5 would ordinarily be *American* but we are going to overlay two graphs, and the label for the first column becomes the label for the X-axis; thus we want the label to read *Price* rather than American.)

To construct the Graph:
   a.   Select **Graph>Histogram**
   b.   Under *Graph Variables*: click in the first box for Graph 1 and move to the box at the left and double-click on the label for column c5, which is *Price* (not the label for column c1, which is *Prices*).
   c.   Click in the next box for Graph 2 under *Graph Variables* and in the box at the left double-click on *NY*.
   d.   Move to the next group of boxes. In order for both the Polygons to have both a symbol and for the symbols to be connected, the table must be filled out as shown below, where the **bold** is what will be selected.

For Display:

Item	Display	For each	Group Variables
1	**Symbol**	**Graph**	
2	**Connect**	**Graph**	

   e.   Title: Click on **Annotation>Title**
      1.   In the box labeled 1 type **Percentage Polygon for American (solid line)**
      2.   In the box labeled 2 (the second line of the title) type **and NY Prices (dashed line)**

3. Then click in the first box under the label *Text Size* and type **1** to reduce the size.

4. Do the same for the next box under *Text Size*. Click on *OK*.

f. Axis tick marks: Click on **Frame>Tick** and you will see a screen with the following headings. The numbers enclosed in boxes are the ones to enter. In Row 1, enter **9** for the Number of Major ticks and enter **4** for the number of Minor Ticks. In Row 2, enter **4** and **4**. (Initially the default values may be *Auto*: just delete these and enter the numbers as shown.) Then Click OK.

	Direction	Side	Positions	Number of Major	Number of Minor
1	X			9	4
2	Y			4	4

g. Click on the Options button:
1. Click on *Percent* for *Type of Histogram* and click on *MidPoint* for *Type of Intervals*.
2. Click on the Radio button *Midpoint/cutpoint positions:*, click in the box to its right, and type **-5:95/10**, which is a shorthand way to indicate the numbers on the X-axis go from -5 to 95, by units of 10.
3. Click on OK.

h. Click on **Frame>Multiple Graphs** and click on *Overlay graphs on the same page*.

i. Click on *OK* twice and the graph will appear.

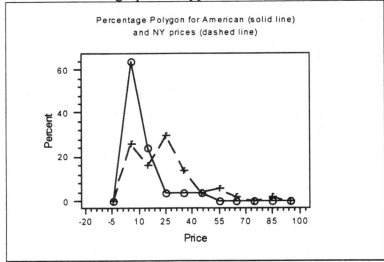

Percentage Polygon for American (solid line) and NY prices (dashed line)

j. If you typed rather than using the menus,
```
MTB > Histogram 'Price' 'NY';
SUBC>    Percent;
SUBC>    MidPoint -5:95/10;
SUBC>    Symbol;
SUBC>    Connect;
SUBC>    Title "Percentage Polygon for American (solid
    line)";
SUBC>      TSize 1;
SUBC>    Title "and NY prices (dashed line)";
SUBC>      TSize 1;
SUBC>    Overlay;
SUBC>    Tick 1;
SUBC>      Number 9 4;
SUBC>    Tick 2;
SUBC>      Number 4 4.
```

Cumulative Percentage Polygon (Ogive):
Follow the same instructions as for the previous problem except for those shown below. (Note that if you have just completed the previous graph, everything will be filled in except the changes shown below.)

a. Suppose you have selected **Graph>Histogram**
b. Title: Click on **Annotation>Title**
    1. In the box labeled 1 add the word *Cumulative* at the beginning, so the title is
       Cumulative Percentage Polygon for American (solid line)
c. The Axis tick marks should still be set for 9 & 4 for X and 4 & 4 for Y. If not, Click on **Frame>Tick** and you will see a screen with the following headings. The numbers enclosed in a box are the ones to enter. In Row 1, enter **9** for the Number of Major ticks and enter **4** for the number of Minor Ticks. In Row 2, enter **4** and **4**. (Initially the default values may be *Auto*: just delete these and enter the numbers as shown.) Then Click OK.

	Direction	Side	Positions	Number of Major	Number of Minor
1	X			9	4
2	Y			4	4

d. Click on the Options button:
    1. Click on *Cumulative Percent* for *Type of Histogram* and click on *MidPoint* for *Type of Intervals*.
    2. The *Midpoint/cutpoint positions:* section should still have the correct interval description in its box: **-5:95/10**, which is a shorthand way to indicate the numbers on the X-axis go from -5 to 95, by units of 10.

3.   Click on OK twice and the graph will appear.

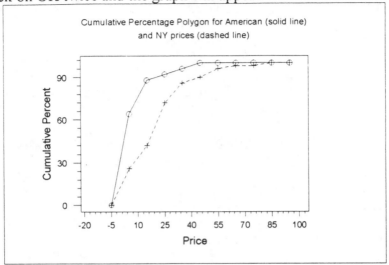

e.   If you typed rather than using the menus,

```
MTB > Histogram 'Price' 'NY';
SUBC>    Cumulative;
SUBC>    Percent;
SUBC>    MidPoint -5:95/10;
SUBC>    Symbol;
SUBC>    Connect;
SUBC>    Title "Cumulative Percentage Polygon for American
     (solid line)";
SUBC>       TSize 1;
SUBC>    Title "and NY prices (dashed line)";
SUBC>       TSize 1;
SUBC>    Overlay;
SUBC>    Tick 1;
SUBC>       Number 9 4;
SUBC>    Tick 2;
SUBC>       Number 4 4.
```

SUMMARIZING AND DESCRIBING NUMERICAL DATA

The goal of this chapter is to describe the characteristics of numerical data, and to develop descriptive summary measures.

PROBLEM 4.5:

If the data aren't currently loaded, open the file by using the menu **File>Other Files>Import ASCII Data**; type **c1** for *column* and click *OK*. Give the file name *flashbat.dat* and its location. Type the heading **Hours** for column c1.

The goal is to compute the Mean, Median, Mode, Midrange, and Midhinge.

a. Select **Stat>Basic Statistics>Descriptive Statistics**
b. Double-click on *Hours*.
c. If you typed rather than using the menus,
   MTB > **Describe 'Hours'**.
d. The display is shown.
```
Descriptive Statistics
Variable N      Mean    Median    TrMean    StDev    SEMean
Hours    13     473.5   451.0     440.2     210.8    58.5

Variable  Min     Max        Q1        Q3
Hours     264.0   1049.0     307.5     553.5
```
e. The Mean is shown above as 473.5, and the Median is 451.0 .
f. The Mode is the number which occurs the most frequently. For a small problem like this you can just observe the numbers and notice that there is no mode, since there is

only one of each number. For a problem with a larger sample size you may type at the session window the following command, which would then list the values and the number of times each occurs, which would make it easy to determine which value occurs the most frequently.

```
MTB > tally 'Hours'
```

g.   The Midrange is the average of the largest and the smallest values:

$$\text{Midrange} = \frac{1049 + 264}{2} = 656.50 \, .$$

h.   The Midhinge is the average of the third and first quartiles:

$$\text{Midhinge} = \frac{553.5 + 307.5}{2} = 430.50 \, .$$

PROBLEM 4.20:

Continuing with the *flashbat.dat* data from Problem 4.5, the goal is to compute the Range, Variance, and the Standard Deviation.

a.   The display below is reproduced from the previous problem. (The *Describe* command was used, either through the menus or through typing at the prompt at the Session Window.)

```
Descriptive Statistics
Variable N      Mean    Median    TrMean    StDev    SEMean
Hours    13     473.5   451.0     440.2     210.8    58.5

Variable  Min      Max       Q1        Q3
Hours     264.0    1049.0    307.5     553.5
```

b.   The Range is the difference between the largest and smallest values.
      Range = 1049 - 264 = 785.

c.   The Standard Deviation is shown as *StDev* = 210.8 = S.

d.   The Variance is (Standard deviation)^2, so Variance = S^2 = (210.8)^2 = 44,436.64 .

PROBLEM 4.37

Continuing with the *flashbat.dat* data from Problem 4.5, the goal is to compute the 5-Number Summary and form the box-and-whiskers plot.

If the data aren't currently loaded, open the file by using the menu **File>Other Files>Import ASCII Data**; type **c1** for *column* and click *OK*. Give the file name *flashbat.dat* and its location. Type the heading **Hours** for column c1. (Equivalently, type **'Hours'** instead of *c1* initially and you don't have to later type the heading.)

a.   Select **Stat>Basic Statistics>Descriptive Statistics**

b.   Double-click on *Hours*.

c.   If you typed rather than using the menus,
      ```
      MTB > Describe 'Hours'.
      ```

d. The display is shown.

```
Descriptive Statistics
Variable   N        Mean     Median     TrMean     StDev    SEMean
Hours      13      473.5      451.0      440.2     210.8      58.5

Variable   Min       Max        Q1         Q3
Hours     264.0    1049.0     307.5      553.5
```

e. The *Five-Number Summary* consists of
    {Minimum, 1st Quartile, Median, 3rd Quartile, Maximum}
    1. From the Minitab display shown above,
        {Minimum, 1st Quartile, Median, 3rd Quartile, Maximum} =
        { 264, 307.5, 451, 553.5, 1049}

f. To form the box-and-whiskers plot:
    1. From the menu select **Graph>Boxplot**
    2. For *Graph Variables* click in the box under *Y* and double-click *Hours* in the box at the left.
    3. Select **Annotation>Title** and type the title **Boxplot for Hours of Battery Life**
    4. Click in the box under *Text Size* and type **1** to reduce the title size and click *OK*.
    5. Click on the *Options* button and click on *Transpose X and Y*. Click OK.
    6. Axis tick marks: Click on **Frame>Tick** and you will see a screen with the following headings. The numbers enclosed in a box are the ones to enter. Ignore Row 1. In Row 2, enter **6** for the Number of Major ticks and enter **9** for the number of Minor Ticks. (Initially the default values may be *Auto*: just delete these and enter the numbers as shown.) Then Click OK twice.

	Direction	Side	Positions	Number of Major	Number of Minor
1	X			Auto	Auto
2	Y			6	9

g. The box-and-whiskers plot is shown below. Note that the data are right-skewed.

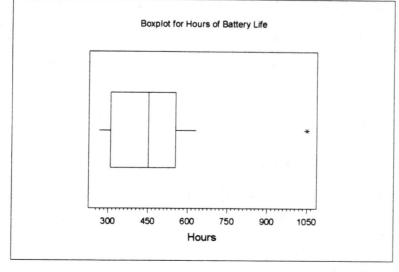

Boxplot for Hours of Battery Life

h.   If you typed rather than using the menus,
```
MTB > Boxplot 'Hours';
SUBC>    Transpose;
SUBC>    Box;
SUBC>    Symbol;
SUBC>      Outlier;
SUBC>    Title "Boxplot for Hours of Battery Life";
SUBC>      TSize 1;
SUBC>    Tick 1;
SUBC>    Tick 2;
SUBC>      Number 6 9.
```

Data From a Population

PROBLEM 4.48
If the data aren't currently loaded, open the file by using the menu **File>Other Files>Import ASCII Data**; type **c1** for *column* and click *OK*. Give the file name *cancer.dat* and its location. Type the heading **Cancer** for column c1.

The goal is to compute the mean, median, mode, midrange, midhinge, range, interquartile range, variance, standard deviation, and the coefficient of variation. In addition, form the box-and-whisker plot.

(Note that this data set (*Cancer*) is considered a *population* rather than a sample from a population. *Minitab* treats this population data as if it were sample data, so that for the standard deviation it uses 49 (n-1=50-1=49) as the divisor rather than the correct 50. Even with this limitation, it is still easier to allow *Minitab* to analyze the data and then to do a separate computation for the standard deviation to obtain the correct population value.)

a.   Select **Stat>Basic Statistics>Descriptive Statistics** for the column *Cancer*.
b.   The display is shown (with an incorrect *StDev* which will be corrected later).
```
Descriptive Statistics
Variable         N      Mean    Median    TrMean     StDev    SEMean
Cancer          50    389.98    397.00    393.07     56.83      8.04

Variable        Min       Max        Q1        Q3
Cancer       229.00    500.00    363.00    435.00
```
c.   If you typed rather than using the menus,
```
MTB > Describe 'Cancer'.
```
d.   The Mean is shown above as 389.98, and the Median is 397 .
e.   The Mode is the number which occurs the most frequently. For the mode, at the Session Window type the command shown below. You can notice that all values appear once except that 5 values appear twice. Therefore, there is no one value which appears the most, and so there is no single mode.
```
MTB > tally 'Cancer'
```
f.   The Midrange is the average of the largest and the smallest values:

$$\text{Midrange} = \frac{500 + 229}{2} = 364.50 .$$

g. The Midhinge is the average of the third and first quartiles:
$$\text{Midhinge} = \frac{435 + 363}{2} = 399 .$$

h. The Range is the difference between the largest and smallest values.
Range = 500 - 229 = 271 .

i. The Standard Deviation is shown as *StDev* = 56.83 = S. *Recall that this is the standard deviation if the data set were a sample.* Since it is a population, we can easily obtain the correct *StDev* in several ways.

1. There are several formulas for computing the variance and standard deviation for a population, and these could easily be used at the Session Window in *Minitab*.

2. One formula is $\sigma_x^2 = \dfrac{\Sigma(X^2) - \dfrac{(\Sigma X)^2}{N}}{N}$, which may be computed in *Minitab* using the *SSQ* function for computing the sum of squared values such as $\Sigma(X^2)$, and using *SUM* for computing the sum ΣX. Raising to a power in *Minitab* is accomplished using "**", such as **2 to square a value. Taking the square root employs the *SQRT* function. In the computation below we assign the population variance to the variable k1, and the standard deviation to the variable k2, and then both k1 and k2 are displayed using the *print* command.

```
MTB > let k1 = (SSQ(c1)-(SUM(c1)**2)/50)/50
MTB > let k2=SQRT(k1)
MTB > print k1-k2
```

a. The resulting display is shown below.
```
Data Display
K1          3165.18
K2          56.2599
```

3. Therefore the variance of the population is 3,165.18, and the standard deviation of the population is 56.26.

j. The variance of the population was computed above: 3,165.18. Of course the variance is the square of the standard deviation, so
$$\sigma_x^2 = \sigma_x * \sigma_x = 56.2599 * 56.2599 = 3,165.18 .$$

k. The interquartile range is 3rd Quartile - 1st Quartile = 435 - 363 = IQR = 72 .

l. *Five-Number Summary*: From the Minitab display shown above,
{Minimum, 1st Quartile, Median, 3rd Quartile, Maximum} =
{ 229, 363, 397, 435, 500}

m. To form the box-and-whiskers plot:

1. From the menu select **Graph>Boxplot**

2. For *Graph Variables* click in the box under *Y* and double-click *Cancer* in the box at the left.

3. Select **Annotation>Title** and type the title **Boxplot for Cancer incidence, Ex 4.48**

4. Click in the box under *Text Size* and type **1** to reduce the title size and click *OK*.

5.   Click on the *Options* button and click on *Transpose X and Y*. Click OK.

6.   Axis tick marks: Click on **Frame>Tick** and you will see a screen with the headings shown below. The numbers enclosed in a box are the ones to enter. Ignore Row 1. In Row 2, enter **6** for the Number of Major ticks and enter **4** for the number of Minor Ticks.   (Initially the default values may be *Auto*: just delete these and enter the numbers as shown.) Then Click OK twice.

	Direction	Side	Positions	Number of Major	Number of Minor
1	X			Auto	Auto
2	Y			6	4

n.   The box-and-whiskers plot is shown below. Note that the data are left-skewed.

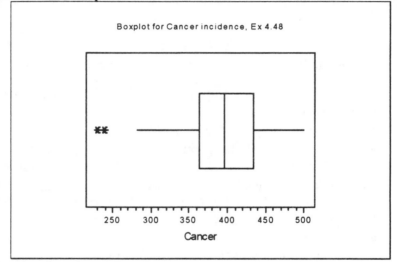

o.   If you typed rather than using the menu,

```
MTB > Boxplot 'Cancer';
SUBC>    Transpose;
SUBC>    Box;
SUBC>    Symbol;
SUBC>      Outlier;
SUBC>    Title "Boxplot for Cancer incidence, Ex 4.48";
SUBC>      TSize 1;
SUBC>    Tick 1;
SUBC>    Tick 2;
SUBC>      Number 6 4.
```

PRESENTING CATEGORICAL DATA IN TABLES AND CHARTS

The goal of this chapter is to show how to organize and present categorical data in the form of tables and charts.

PROBLEM 5.1

There were 616 responses to a poll, and the data are as follows, in response to the question "Should the cooperative hire a supervisor?"

Yes = 146; *No* = 91; *Not sure* = 58; *No response* = 321

Type the heading **Response** for column c1 (when the alpha data are entered into the column the designation will become C1-A for Alpha instead of C1). Type the heading **Freq** in c2.

      a.   (Note: If the arrow at the far left of the Data Window is pointing to the right, click on that arrow so it points down. This causes the Enter key to move down rather than moving right.) In column c1 (which will become C1-A when you enter the text labels) under the heading type in the first four rows the labels **(No, Yes, Not sure, No response)**. In column c2 under the heading type the frequencies for each response **(146, 91, 58, 321)**.

b.   The worksheet should look similar to the following. (Note the arrow at the left which is pointing down.)

↓	C1-A	C2
	Response	Freq
1	Yes	146
2	No	91
3	Not sure	58
4	No response	321

Bar Graph
Next construct a bar graph for the responses.

There are two approaches for obtaining the bar graph.

a.   The obvious first choice (which we will not pursue) is to use **Graph>Histogram**, but this would require some extra effort on our part. The *Histogram* command is appropriate only when you have a column with the individual responses: for instance if column c3 contained 146 1's (which stand for the "Yes" response, 91 2's (for the "No" response), 58 3's (for "Not sure") and 321 4's (for "No response"), then you could use the **Graph>Histogram** approach. Instead of a column with the individual responses, we have the summary: column c2 contains the frequencies or counts of each response.

    1.   Here is how you would obtain the bar graph using the **Graph>Histogram** approach. Type the **bolded** portion:

```
MTB > set c3
DATA> 146(1)  91(2)  58(3)  321(4)
DATA> end
```

    2.   This would give you 616 entries in column c3.

    3.   Then select **Graph>Histogram** and for *Graph Variables:* choose *c3* and click on *OK*.

b.   An easier method (which we'll pursue) is to use **Graph>Chart**. This has several advantages: we don't need to build the column of individual values; and the category names on the X-axis are the natural names ("No" and "Yes" and so on instead of "1" and so on).

Obtaining the bar graph using the *Chart* approach.

a.   Select **Graph>Chart**

b.   For the *Graph Variables* section: Under *Function* type **sum** or select it from drop-down menu. Click in the box under *Y* and double-click on the column name *Freq* in the box to the left. Click in the box under *X* and double-click on the column name *Response* in the box to the left. This section will look similar to the box below.

For Graph Variables:	Graph	Function	Y	X
	1	**sum**	**Freq**	**Response**
	2			

c.   For the next section, it should by default have the entries shown below; if not, change them to conform to this structure.

For Display:

Item	Display	For each	Group Variables
1	**Bar**	**Graph**	
2			

d.   Title: Click on **Annotation>Title** and type **Histogram of Responses (Ex 5.1)** Then click in the box under Text size and type **1** to reduce the title size. Click *OK*.

e.   Axis tick marks: Click on **Frame>Tick** and you will see a screen with the headings shown below. The values enclosed in a box are the ones to enter. In Row 1 (for the X-axis), type *Auto* (unless *Auto* is already there) for the Number of Major ticks and for the number of Minor ticks. In Row 2 (for the Y-axis), enter 7 and **4**. (Initially the default values may be *Auto*: just delete these and enter the numbers as shown. We've selected 7 so that the numbers will start at 50.) Then Click OK.

	Direction	Side	Positions	Number of Major	Number of Minor
1	X			Auto	Auto
2	Y			7	4

f.   Select **Frame>Min and Max** and click on the check box to the left of *Y minimum* and then click in the box to the right of *Y minimum* and type **0**. This will include zero in the bar graph. Click on *OK*.

g.   Select **Frame>Grid** and click on the arrow to the right of *Direction* and click on *Y*. The rest of the row will be filled in with the default of a dotted line, and this will be all right. (We've included the grid just for your information. It may or may not make the graph easier to read.) Click on *OK*.

h.   Click *OK* and the graph will appear.

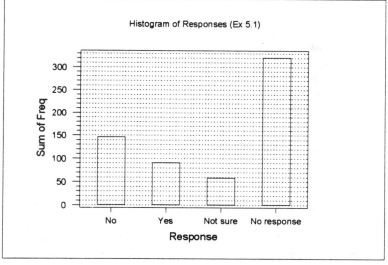

Pie Graph

Construct a pie graph for percentage.

a.   In column c4 enter the heading **PieSlice** and type these values in the first four rows: **2, 3, 5**, and **6**, which will provide hatching for the pie chart.

b.   Select **Graph>Pie Chart**

c.   Click on *Chart Table*; click in the box by *Categories in* and double-click on *Response;* click in the box by *Frequencies in* and double-click on *Freq*.

d.   Click in the box by *Explode slice number(s)* and type *3* so that the "Not sure" slice will be exploded.

e.   Click in the box by *Title* and type **Pie chart percentage for response (Ex 5.1)**

f.   Click on the *Options...* button: (1) Click on the choice *Add lines connecting labels to slices*; (2) under the heading *Fill types* click on *Use types in* and click in the box next to *Use types in* and then double-click on *PieSlice* in the box at the far left, which will provide the hatching; (3) click *OK*.

g.   Click *OK* and the graph shown below will appear.

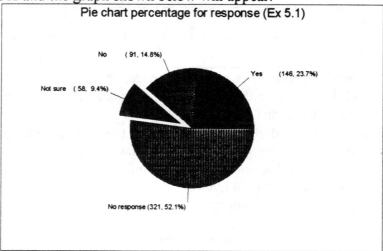

h.   If you typed rather than using the menu,

```
MTB > %Pie 'Response';
SUBC>    Counts 'Freq';
SUBC>    Explode 3;
SUBC>     Title 'Pie chart percentage for response (Ex
     5.1)';
SUBC>    Lines;
SUBC>    Types 'PieSlice'.
```

PARETO CHART

Construct a Pareto Chart for the response to the hiring question for problem 5.1. The Pareto Chart provides a convenient structure for identifying the most important or significant classes in a group. The table below consists of two separate but related graphs: (1) a bar chart indicating in order which of the responses had the highest percentage. For instance, the category *No response* has the largest bar, because this category had the highest

percentage. The other answers are also indicated in order of percentage. (2) A cumulative polygon which shows the cumulative percentage of the responses.
a.   Select **Stat>SPC>Pareto Chart**
b.   Click on *Chart defects* table;
       Click in box by *Labels in* and double-click on *Response;* click in box by *Frequencies in* and double-click on *Freq.*
c.   Click in box by *Title* and type **Pareto Chart for Hiring Response (Ref Ex 5.1)**
d.   Click on the *OK* button to obtain the chart shown below.

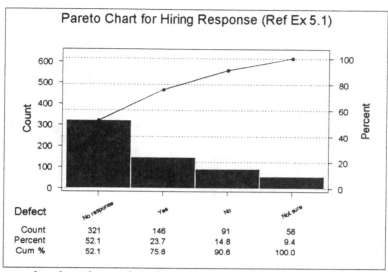

Defect	No response	Yes	No	Not sure
Count	321	146	91	58
Percent	52.1	23.7	14.8	9.4
Cum %	52.1	75.8	90.6	100.0

e.   If you typed rather than using the menu,
```
MTB >  %Pareto 'Response';
SUBC>    Counts 'Freq';
SUBC>    Others 95;
SUBC>    Title 'Pareto Chart for Hiring Response (Ref Ex 5.1)'.
```

CONTINGENCY TABLE
The following data pertain to information on employees.

If the data aren't currently loaded, open the file by using the menu **File>Other Files>Import ASCII Data**; type **c1-c29** for *column* and click *OK*. Give the file name *empsat.dat* and its location. Type the heading **Occup** for column c3 and **Gender** for column c6.
a.   Select **Stat>Tables>Cross Tabulation**
b.   Click in *Classification variables*;
       double-click on *Gender*; double-click on *Occup*
c.   Under *Display*, check all four boxes: *Counts; Row percents; Column percents; Total percents.*
d.   Click on the *OK* button to obtain the contingency table shown below. (The headings in **bold** were added later for clarification, as were the heavy borders.)

```
Tabulated Statistics
ROWS: Gender       COLUMNS: Occup
```

		MGL	PROF	TEC/SAL	ADMSPT	SERV	PROD	LABOR	
		1	2	3	4	5	6	7	ALL
Counts Males	1	36	33	34	14	18	51	47	233
Row Pct		15.45	14.16	14.59	6.01	7.73	21.89	20.17	100
Col Pct		55.38	50	59.65	21.54	62.07	94.44	73.44	58.25
Percent		9	8.25	8.5	3.5	4.5	12.75	11.75	58.25
Counts Females	2	29	33	23	51	11	3	17	167
Row Pct		17.37	19.76	13.77	30.54	6.59	1.8	10.18	100
Col Pct		44.62	50	40.35	78.46	37.93	5.56	26.56	41.75
Percent		7.25	8.25	5.75	12.75	2.75	0.75	4.25	41.75
Counts	ALL	65	66	57	65	29	54	64	400
Row Pct		16.25	16.5	14.25	16.25	7.25	13.5	16	100
Col Pct		100	100	100	100	100	100	100	100
Percent		16.25	16.5	14.25	16.25	7.25	13.5	16	100

```
CELL CONTENTS --
            COUNT
            % OF ROW
            % OF COL
            % OF TBL
```

e.   If you typed rather than using the menu,
```
MTB > Table  'Gender' 'Occup';
SUBC>    Counts;
SUBC>    RowPercents;
SUBC>    ColPercents;
SUBC>    TotPercents.
```

SUPER TABLE

Use the same procedures as before, but this time use the headings **Lottery** and **Occup**. Place the heading **Lottery** in column c11. The heading *Lottery* pertains to whether one would quit work after winning a lottery.

Tabulated Statistics
ROWS: Gender   COLUMNS: Occup

			MGL	PROF	TEC/SAL	ADMSPT	SERV	PROD	LABOR	
			1	2	3	4	5	6	7	ALL
Counts	Males	1	36	33	34	14	18	51	47	233
Row Pct			15.45	14.16	14.59	6.01	7.73	21.89	20.17	100
Col Pct			55.38	50	59.65	21.54	62.07	94.44	73.44	58.25
Percent			9	8.25	8.5	3.5	4.5	12.75	11.75	58.25
Counts	Females	2	29	33	23	51	11	3	17	167
Row Pct			17.37	19.76	13.77	30.54	6.59	1.8	10.18	100
Col Pct			44.62	50	40.35	78.46	37.93	5.56	26.56	41.75
Percent			7.25	8.25	5.75	12.75	2.75	0.75	4.25	41.75
Counts		ALL	65	66	57	65	29	54	64	400
Row Pct			16.25	16.5	14.25	16.25	7.25	13.5	16	100
Col Pct			100	100	100	100	100	100	100	100
Percent			16.25	16.5	14.25	16.25	7.25	13.5	16	100

ROWS: Lottery   COLUMNS: Occup

			MGL	PROF	TEC/SAL	ADMSPT	SERV	PROD	LABOR	
			1	2	3	4	5	6	7	ALL
Counts	Stop Work	1	45	44	35	38	20	42	33	257
Row Pct			17.51	17.12	13.62	14.79	7.78	16.34	12.84	100
Col Pct			69.23	66.67	61.4	58.46	68.97	77.78	51.56	64.25
Percent			11.25	11	8.75	9.5	5	10.5	8.25	64.25
Counts	Don't Stop	2	20	20	21	27	9	12	30	139
Row Pct			14.39	14.39	15.11	19.42	6.47	8.63	21.58	100
Col Pct			30.77	30.3	36.84	41.54	31.03	22.22	46.88	34.75
Percent			5	5	5.25	6.75	2.25	3	7.5	34.75
Counts	Not Sure	3	0	2	1	0	0	0	1	4
Row Pct			--	50	25	--	--	--	25	100
Col Pct			--	3.03	1.75	--	--	--	1.56	1
Percent			--	0.5	0.25	--	--	--	0.25	1
Counts		ALL	65	66	57	65	29	54	64	400
Row Pct			16.25	16.5	14.25	16.25	7.25	13.5	16	100
Col Pct			100	100	100	100	100	100	100	100
Percent			16.25	16.5	14.25	16.25	7.25	13.5	16	100

CELL CONTENTS --
 COUNT
  % OF ROW
  % OF COL
  % OF TBL

a. If you typed rather than using the menu,

```
MTB > Table 'Lottery' 'Occup';
SUBC>   Counts;
SUBC>   RowPercents;
SUBC>   ColPercents;
SUBC>   TotPercents.
```

BASIC PROBABILITY

The goal of this chapter is to develop an understanding of the basic concepts of probability, which will be needed for the study of probability distributions.

PROBLEM 6.66:

Is there a relationship between the ownership of foreign-made cars and the geographic area? A survey indicated the following counts of people falling into various groups. The table below is a table of *counts*. For convenience, we'll also convert it into a table of *probabilities*, which often provides an easier platform for answering numerous probability questions.

Table of *counts*

	AREA TYPE			
	Large City	Suburb	Rural	TOTALS
Own foreign car	90	60	25	175
Don't own foreign	110	90	125	325
TOTALS	200	150	150	500

1. In order for Minitab to provide the table of *probabilities*, the data must be put into a certain form.
    a. Label one column **Freq**, the next one **Foreign**, and the third one **Area**.
    b. In the *Freq* column, enter the 6 values: 90, 60, 25, 110, 90, 125. In the *Foreign* column, enter a 1 if the corresponding entry in Freq pertains to the owner of a foreign

car, or enter 2 if it is a nonowner. In the *Area* column, enter a 1 if the corresponding entry in Freq pertains to *large city*, enter a 2 if it pertains to *suburb*, and enter a 3 if it pertains to *rural*.

c.   In other words, the 90 people who own foreign cars and are from a large city all have a 1 for *Foreign* and a 1 for *area*. The 125 who don't own foreign cars and are from a rural area all have a 2 for *foreign* and a 3 for *rural*.

d.   Therefore, the Minitab worksheet should look like this:

Freq	Foreign	Area
90	1	1
110	2	1
60	1	2
90	2	2
25	1	3
125	2	3

2.   Now to obtain the probabilities,
   a.   **Select Stat>Tables>Cross Tabulation**
   b.   Click in the box for *Classification variables*; double-click on *Foreign*; double-click on *Area*
   c.   Under *Display*, check the box *Total percents*.
   d.   Click in the box *Frequencies are in:* and double-click *Freq*.
   e.   Click *OK* to obtain the contingency table shown below

```
Tabulated Statistics
ROWS: Foreign       COLUMNS: Area
1         2         3         ALL

  1     18.00     12.00      5.00     35.00
  2     22.00     18.00     25.00     65.00
ALL     40.00     30.00     30.00    100.00

    CELL CONTENTS --
              % OF TBL
```

   f.   If you typed rather than using the menu,
```
MTB > Table  'Foreign' 'Area';
SUBC>    Frequencies 'Freq';
SUBC>    TotPercents.
```

For convenience, the results are displayed again in the table below, but they are shown as *probabilities* rather than *percentages*.

table of *probabilities*	AREA TYPE			
	Large City	Suburb	Rural	TOTALS
Own foreign car	.18	.12	.05	.35
Don't own foreign	.22	.18	.25	.65
TOTALS	.40	.30	.30	1.00

With the probabilities in this form, the questions of problem 6.66 can easily be answered.
a.   P(foreign) = .35, from the table (or 175/500 from the table of counts)
b.   P(suburb) = .30, from the table (or 150/500 from the table of counts)
c.   P(foreign OR large city) = P(foreign) + P(large city) - P(foreign ∩ large city)
     = .35 + .40 - .18 = .57. (Subtracting because of double-counting.)
     1.   Or, just add the individual probabilities: (.18 = foreign ∩ large) + (.12 = foreign ∩ suburb) + (.05 = foreign ∩ rural) + (.22 = not foreign ∩ large) = .57 .
d.   P(large city OR suburb) = P(large city) + P(suburb) - P(large city ∩ suburb)
     = .40 + .30 - 0 (since one doesn't live in both at the same time) = .70 .
e.   P(large city AND owns foreign) = .18, the intersection in the body of the table, at the intersection of large city AND owns a foreign car. (Or, 90/500 from the table of counts.)
f.   P(rural OR not foreign car) = P(rural) + P(not foreign car) - P(rural ∩ not foreign car)
     = .30 + .65 - .25 = .70 .
g.   We know the person lives in a suburb: what is the probability the person owns a foreign car?   This conditional probability is P(foreign|suburb) = $\frac{P(foreign \cap suburb)}{P(suburb)} = \frac{.12}{.30} = 0.40$.

(From the table of counts, this would be $\frac{60/500}{150/500} = \frac{60}{150} = 0.40$.)

h.   There are several forms of a test to determine whether the area is independent of foreign-car ownership. In general, if knowing the area provides a different probability that a person owns a foreign car than when the area is not known, then the factors are not independent. That is, if any particular person from the suburbs has the same chance of owning a foreign car as just any particular person regardless of the area, then the area doesn't make any difference, and the factors are independent.

Test for independence: is P(foreign|suburb) = P(foreign)? From before,
[P(foreign|suburb) = .40] ≠ [P(foreign) = .35], so knowing which area the person lives in makes a difference, and therefore the factors are not statistically independent.

Generating a distribution:
The following simple set of commands shows how to generate a set of random variates. (In Minitab, there are a number of choices of the underlying distribution besides normal.)
a.   First, generate 1,000 normal variates. (The example uses column C1, but pick any column.) At the Session Window type
```
MTB > 1000 random c1;      #note the semicolon
SUBC> normal.             #note the period
```
b.   Type the label for c1: **Z-Score**
c.   Select **Graph>Histogram**
d.   For Graph Variables:, choose *Z-Score*
e.   For *Display*, choose *Bar*, and for *For each*, choose *Graph*.

f. Title: Click on **Annotation>Title** and type **1,000 normal variates, N(0,1)**. Click *OK*.

g Click on the Options button, and click on *frequency* for *Type of Histogram*. Click *OK*.

h. Click on *OK* to obtain the graph.

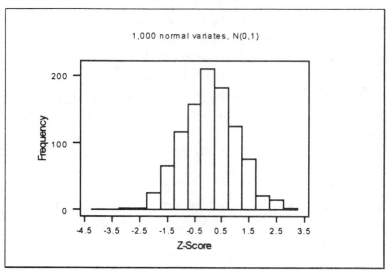

i. If you typed rather than using the menu,

```
MTB > Histogram 'Z-Score';
SUBC>    MidPoint;
SUBC>    Bar;
SUBC>    Title "1,000 normal variates, N(0,1)";
SUBC>     TSize 1;
SUBC>    Axis 1;
SUBC>    Axis 2.
```

SOME IMPORTANT DISCRETE PROBABILITY DISTRIBUTIONS

The goal of this chapter is to describe the concept of mathematical expectation and its applications, and to show the role of mathematical models in decision analysis.

Problem 7.1:
- a. Type the headings: for c1 type **X**, for c2 type **P(X**), and for c3 type **X*P(X).**
- b. Type the data of problem 7.1 into c1 and c2.
- c. For c3, compute the products of X*P(X): Type at the Session Window
    ```
    MTB> let 'X*P(X)' = 'X' * 'P(X)'
    ```
- d. Compute the mean, which is the sum of the products in c3, or

$$\mu_x = E(X) = \sum_{i=1}^{N} X_i * P(X_i) \text{, and store it in the constant K1: type}$$

    ```
    MTB> let k1=sum(c3)
    ```
- 1. Note that for mathematical functions (including SUM and SQRT used in this section), the parentheses are necessary.

e.   Compute the variance of the probability distribution, by $\sigma_x^2 = \sum_{i=1}^{N}(X_i - \mu_x)^2 * P(X_i)$ and

store it in the constant K2: type
```
MTB> let k2=sum((c1-k1)**2*c2)
```

f.   Compute the standard deviation σ_x and store it in K3: type
```
MTB> let k3 = sqrt(k2)
```

g.   Print the results to the Session Window, type
```
MTB> print c1-c3, k1-k3
```

h.   The result is shown below. Note that the headings shown below in italics, *Mean*, *Variance*, and *Standard Deviation*, were typed directly into the Session Window for the convenience of the student, and they don't appear on the Minitab display.

```
Data Display
```

K1	1.00000	*Mean*
K2	1.50000	*Variance*
K3	1.22474	*Standard Deviation*

Row	X	P(X)	X*P(X)
1	0	0.50	0.0
2	1	0.20	0.2
3	2	0.15	0.3
4	3	0.10	0.3
5	4	0.05	0.2

Questions

Various probability questions could be posed and answered from the output shown above.

a.   What is the probability that X will be 2? From the table above, $P(X=2) = 0.3$.

b.   What is the probability that X will be > 2? From the table above,
$P(X>2) = [P(X=3) = .10] + [P(X=4) = .05] = 0.15$.

A plot may be obtained by:

a.   Select **Graph>Plot**

b.   For *Graph Variables*: for *Y* double-click on *P(X)*; for X double-click on *X*.

c.   For *Display*, under Item 1 choose *Symbol*, and under *For each*, choose *Graph*.
Then for Item 2 under *Display* either type *Project* or select it from the drop down list, and under *For each* choose or type *Graph*. The result is shown below.

For Display:	Item	Display		For each
	1	**Symbol**		**Graph**
	2	**Project**		**Graph**

d.   Click on *Symbol* and click on the *Edit Attributes* button and under *Type* click on the arrow and select *Solid circle*, and click *OK*.

e.   Title: Select **Annotation>Title** and type **Distribution A: (Ex 7.1)**
Then click in the first box under the label *Text Size* and type **1** to reduce the title size.

f.   Click on *OK* twice to obtain the graph.

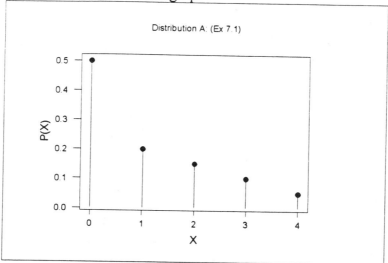

g.   If you typed rather than using the menu,

```
MTB > Plot 'P(X)'*'X';
SUBC>    Symbol;
SUBC>     Type 6;
SUBC>    Project;
SUBC>    Title "Distribution A: (Ex 7.1)";
SUBC>     TSize 1.
```

Distribution B

Using the same procedure, place the data for Distribution B of problem 7.1 into some columns and obtain the results shown below.

a.   Comparing the two distributions: they have different means (1.00 for Distribution A and 3.00 for Distribution B) but the same standard deviation of 1.22.

```
Data Display
K1          3.00000        Mean
K2          1.50000        Variance
K3          1.22474        Standard Deviation

Row    Y    P(Y)    Y*P(Y)
  1    0    0.05     0.0
  2    1    0.10     0.1
  3    2    0.15     0.3
  4    3    0.20     0.6
  5    4    0.50     2.0
```

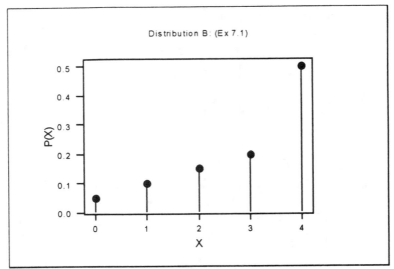

b. If you typed rather than using the menu,
```
MTB > Plot 'P(X)'*'X';
SUBC>    Symbol;
SUBC>      Type 6;
SUBC>    Project;
SUBC>    Title "Distribution B: (Ex 7.1)";
SUBC>      TSize 1.
```

BINOMIAL

Problem 7.30
Suppose 15% of the bills are incorrect. A sample of 3 is selected. (Therefore, p = .15 and n=3.)

a. Type at the Session Window prompt
```
MTB > pdf;                      #including the semicolon
SUBC> binomial n 3 p .15 .   #including the last period
```

b. The result is shown below. This is the binomial probability distribution for a sample of n=3 and with a probability of "success" of .15 . That is, out of the sample of 3, there is a .6141 probability that the number incorrect is 0. Similarly, out of the sample of 3, there is a .0034 probability that the number incorrect is 3.
```
Probability Density Function
Binomial with n = 3 and p = 0.150000
        x        P( X = x)
        0         0.6141
        1         0.3251
        2         0.0574
        3         0.0034
```

The displayed binomial probability distribution may be used to answer the questions.

a. The probability that exactly two bills are incorrect out of 3 is shown in the probability distribution: P(X=2) = .0574.

b.   The probability that no more than two bills are incorrect out of 3 must be computed:
      P(X ≤ 2) = [P(X=0) = .6141] + [P(X=1) = .3251] + [P(X=2) = .0574] = .9966.

c.   The probability that at least two bills are incorrect:
      P(X ≥ 2) = [P(X=2) = .0574] + [P(X=3) = .0034] = .0608.

Plot

a.   In order to obtain a plot, the results have to be stored in columns in the worksheet rather than displayed: first type the heading for c1, **X**, and the heading for c2, **P(X=x)**.

b.   Enter the possible values of X into c1 by typing at the Session prompt
```
MTB > set 'X'
DATA> 0:3
DATA> end.                          #remember the final period.
```

c.   Type at the prompt
```
MTB > pdf c1 c2;                    #including the semicolon
SUBC> binomial n 3 p .15 .         #Note the final period.
```

d.   This places the probability distribution in the X and P(X) columns rather than displaying it at the Session Window.

To obtain a plot:

a.   Select **Graph>Plot**

b.   For Graph Variables: for Y double-click on *P(X=x)*; for X double-click on *X*.

c.   For Display, choose for Item 1 *Symbol*, and for *For each*, choose *Graph*. Then under Item 2 choose *Project* and for *For each*, choose *Point*. This will provide projection lines to the points.

d.   Click on Symbol and click on the *Edit Attributes* button and under *Type* click on the arrow and select *Solid circle*, and click *OK*.

e.   Title: Click on **Annotation>Title** and type **Binomial: Incorrect Bills (Ex 7.30)**. Then click in the box under the label *Text Size* and type **1** to reduce the title size.

f.   Click on *OK* twice to obtain the graph.

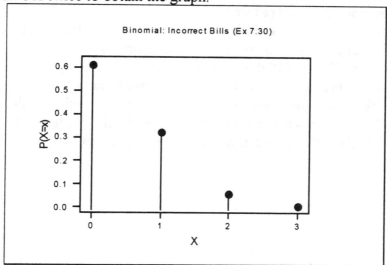

g.   If you typed rather than using the menu,

```
MTB > Plot 'P(X=x)'*'X';
SUBC>    Symbol;
SUBC>     Type 6;
SUBC>    Project;
SUBC>    Title "Binomial: Incorrect Bills (Ex 7.30)";
SUBC>     TSize 1.
```

CDF

Note that the questions could also be answered if rather than the probability distribution, which provides P(X=x), we had the cumulative probability distribution, which provides P(X ≤ x).

a.   To obtain the CDF, type at the Session Window

```
MTB > cdf;                    #including the semicolon
SUBC> binomial n 3 p .15 .  #Remember the final period
```

b.   The result is displayed as

```
Cumulative Distribution Function
Binomial with n = 3 and p = 0.150000
        x        P( X <= x)
        0         0.6141
        1         0.9392
        2         0.9966
        3         1.0000
```

c.   Then the answers are:
1.   P(X=2) = .9966 - .9392 = .0574
2.   P(X ≤ 2) = .9966 directly from the display.
3.   P(X ≥ 2) = 1 - P(X ≤ 1) = 1 - .9392 = .0608.

CDF Plot

a.   CDF plot: Using the same approach as previously described, you could place the CDF data into columns with labels X and $P(X \le x)$. (See the commands after the graph below as a reminder of how to get the CDF data into columns c1 and c2 (or whichever columns you are using).

b.   Important: one additional task is required, which if not employed will lead to a misleading graph which doesn't include the origin.

c.   After selecting **Graph>Plot** as described for the previous problem, and setting up the title and the other tasks, select **Frame>Min and Max**.
1.   Click on the check box to the left of *Y minimum* and then click in the box to the right of *Y minimum* and type **0** to insure that the origin will be part of the graph. Click *OK* and proceed as for the previous problem.

d.   The CDF plot is as shown below.

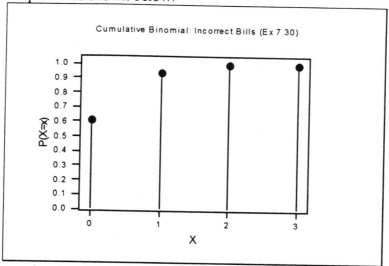

e.   If you typed rather than using the menu,

```
MTB > set 'X'
DATA> 0:3
DATA> end.
MTB > cdf c1 c2;
SUBC> binomial n 3 p .15 .
MTB > Plot 'P(X=x)'*'X';
SUBC>    Symbol;
SUBC>       Type 6;
SUBC>    Project;
SUBC>    Title "Cumulative Binomial: Incorrect Bills (Ex 7.30)";
SUBC>       TSize 1;
SUBC>    Minimum 2 0;
SUBC>    Tick 1;
SUBC>    Tick 2.
```

POISSON DISTRIBUTION

PROBLEM 7.21:

The number of claims per hour to an insurance company is 3.1 on average. The parameter required for the Poisson distribution is the mean = variance = $\lambda = 3.1$.

First, we wish to look at the Poisson distribution for this problem.

a.   Type at the Session Window:

```
MTB > pdf;                    #include semi-colon
SUBC> poisson mean 3.1 .      #include final period
```

b. The result is shown below. This is the Poisson probability density function with mean = variance = $\lambda = 3.1$.

```
Probability Density Function
Poisson with mu = 3.10000
       x          P( X = x)
    0.00            0.0450
    1.00            0.1397
    2.00            0.2165
    3.00            0.2237
    4.00            0.1733
    5.00            0.1075
    6.00            0.0555
    7.00            0.0246
    8.00            0.0095
    9.00            0.0033
   10.00            0.0010
   11.00            0.0003
   12.00            0.0001
   13.00            0.0000
```

c. The output contains probabilities for 13 possible outcomes, although there is theoretically no limit. In this example all probabilities beyond 12 have probability 0.

d. Note that to answer questions it may be convenient to obtain the cumulative distribution function (CDF).

1. Type at the Session Window:

```
MTB > cdf 'X';          #include semi-colon
SUBC> poisson mean 3.1. #include final period
```

2. The output is shown below

```
Cumulative Distribution Function
Poisson with mu = 3.10000
       x          P( X <= x)
    0.00            0.0450
    1.00            0.1847
    2.00            0.4012
    3.00            0.6248
    4.00            0.7982
    5.00            0.9057
    6.00            0.9612
    7.00            0.9858
    8.00            0.9953
    9.00            0.9986
   10.00            0.9996
   11.00            0.9999
   12.00            1.0000
```

Questions

The displayed Poisson probability distribution may be used to answer the questions.

a. The probability that fewer than 3 claims will be made is

$P(X < 3) = [P(X=0) = 0.0450] + [P(X=1) = 0.1397] + [P(X=2) = 0.2165] = 0.4012$.

1. Note that this may be read directly from the CDF output for $P(X \le 2)$.

b.   The probability of exactly 3 claims may be read directly from the PDF output:
      P(X=3) = 0.2237 .

c.   Compute the probability that there will be 3 or more claims.
      1.   You could compute from the PDF output the $P(X \geq 3) = P(X=3) + P(X=4) +$
            ... to an infinite number of terms, but realistically stopping at P(X=13), by
            simply adding the various probabilities.
      2.   However, the easier way is to compute $P(X \geq 3) = 1 - P(X < 3) =$
            1 - { [P(X=0) = 0.0450] + [P(X=1) = 0.1397] + [P(X=2) = 0.2165] = 0.4012 }
            = 1 - 0.4012 = 0.5988 .
      3.   Another approach using the CDF output is $1 - P(X \leq 2) = 1 - 0.4012 = 0.5988$.

d.   Compute the probability that there will be more than 3 claims. There are several
      approaches.
      1.   Using the PDF compute
            $1 - P(X \leq 3) = 1 - \{0.0450 + 0.1397 + 0.2165 + 0.2237\} =$
            1 - 0.6249 = 0.3751 .
      2.   Using the CDF compute
            $1 - P(X \leq 3) = 1 - 0.6248 = 0.3752$ (which differs from the previous by
            rounding).

Plot

Next we obtain a plot of the Poisson distribution when the mean is $\lambda = 3.1$. As in the binomial
example, the data must be sent to the worksheet rather than displayed in the Session Window.

In order to obtain a plot, the results have to be stored in columns in the worksheet rather than
displayed in the Session Window. We'll store the possible values for X in c1 and the
corresponding probabilities in c2.

a.   First type the heading for c1, **X**, and the heading for c2, **P(X=x)**.
b.   Enter the possible values of X into c1 by typing at the Session prompt
      ```
      MTB > set 'X'
      DATA> 0:15
      DATA> end.                      #remember the final period.
      ```
c.   Type at the prompt
      ```
      MTB > pdf c1 c2;                #including the semicolon
      SUBC> poisson mean 3.1 .        #Note the final period.
      ```
d.   This places the probability distribution in the *X* and *P(X)* columns rather than
      displaying it at the Session Window.

Next obtain the plot of the PDF.

a.   Select **Graph>Plot**
b.   For Graph Variables: for *Y* double-click on *P(X=x)*; for *X* double-click on *X*.
c.   For Display, choose for Item 1 *Symbol*, and for *For each*, choose *Graph*. Then under
      Display for Item 2 choose *Project* and for *For each*, choose *Point*. This will provide
      projection lines to the points.
d.   Click on the word *Symbol* and click on the *Edit Attributes* button and under *Type*
      click on the arrow and select *Solid circle*, and click *OK*.
e.   Title: Click on **Annotation>Title** and type **Poisson: Insurance Claims (Ex 7.21)**.
      Then click in the box under the label *Text Size* and type **1** to reduce the title size.

f.   Click on *OK* twice to obtain the graph.

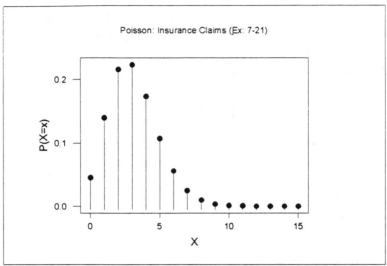

g.   If you typed rather than using the menu (including inserting the data into columns c1 and c2),

```
MTB > set 'X'
DATA> 0:15
DATA> end .

MTB > pdf c1 c2;
SUBC> poisson mean 3.1.
MTB > Plot 'P(X=x)'*'X';
SUBC>    Symbol;
SUBC>      Type 6;
SUBC>    Project;
SUBC>    Title "Poisson: Insurance Claims (Ex: 7-21)";
SUBC>      TSize 1.
```

The Normal Distribution

Prior to working problems, we'll generate a graph of a Standard Normal Distribution, frequently referred to as a N(0,1) distribution. The N(0,1) is a normal distribution with mean of 0 and variance of 1. (The standard deviation is $\sqrt{1} = 1$, as well. When other normal distributions are used which do not have a mean of 0 and a variance of 1, they are automatically transformed to the N(0,1) by using the Z formula.

a. We'll suppose columns c1 and c2 are available. Type the headings **Z** into c1 and **f(Z)** to c2 (or whatever columns you are using).

b. Set a group of Z values into the Z column, by typing in the Session Window the commands shown below.
```
MTB > set 'Z'
DATA> -3.5:3.5/.5        #generates 15 numbers from -3.5 to
     3.5, by .5's
DATA> end.               #include the period
```

c. Place the corresponding normal values into the *f(Z)* column by typing the following commands in the Session Window.
```
MTB > pdf 'z' 'f(z)';
SUBC> normal mu 0 stdev 1 .
```

Plot

The columns *Z* and *f(Z)* have 15 values which determine the N(0,1) distribution. A plot may be generated by using the menus:

a. Select **Graph>Plot**

b. For Graph Variables: for *Y* double-click on *f(Z)*; for *X* double-click on *Z*.

c. For *Display*, choose for Item 1 *Connect*, and for *For each*, choose *Graph*.

d. Title: Click on **Annotation>Title** and type **Standard Normal Distribution: N(0,1)**

e. Click in the first box under *Text Size* and type **1** to reduce the title size. Click *OK*.

f. Click on *OK* to obtain the graph. Note that this graph is not smooth, because there are so few data values.

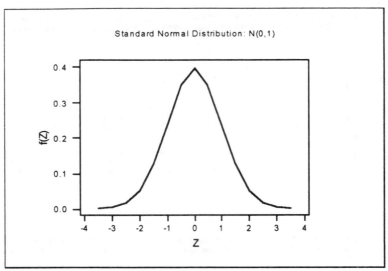

g. If you typed rather than using the menu,
```
MTB > Plot 'f(Z)'*'Z';
SUBC>    Connect;
SUBC>    Title "Standard Normal Distribution: N(0,1)";
SUBC>    TSize 1.
```

Better (Smoother) Graph

a. To obtain a more-attractive graph with 141 values rather than 15, type the following in the Session Window.
```
MTB > set 'Z'
DATA> -3.5:3.5/.05
DATA> end.                    #include the period
```
b. Obtain the *f(Z)* column, as before, by typing
```
MTB > pdf 'Z' 'f(Z)';
SUBC> normal mu 0 stdev 1.
```

Plot

For the new plot, some of the following choices may already be in place; we're adding the tick marks, as well.

a. Select **Graph>Plot**
b. For Graph Variables: for *Y* double-click on *f(Z)*; for *X* double-click on *Z*.
c. For *Display*, choose for Item 1 *Connect*, and for *For each*, choose *Graph*.
d. Title: Click on **Annotation>Title** and type **Standard Normal Distribution: N(0,1): .05 intervals**
e. Click in the first box under *Text Size* and type **1** to reduce the title size. Click *OK*.
f. Axis tick marks: Click on **Frame>Tick** and you will see a screen with the following headings. The numbers enclosed in a box are the ones to enter. In Row 1, enter **15**

for the Number of Major ticks and enter **4** for the number of Minor Ticks.   Then Click OK.

	Direction	Side	Positions	Number of Major	Number of Minor
1	X			15	4

g.   Click on *OK* twice to obtain the graph, which is smoother.

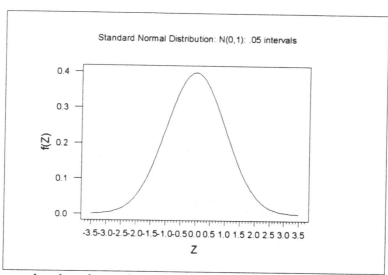

Standard Normal Distribution: N(0,1): .05 intervals

h.   If you typed rather than using the menu,

```
MTB > Plot 'f(Z)'*'Z';
SUBC>    Connect;
SUBC>    Title "Standard Normal Distribution: N(0,1): .05 intervals";
SUBC>     TSize 1.
```

CDF command:

The cdf command in Minitab is used to determine areas from a probability distribution.   In Minitab, for a given or stated z value such as z=1, for instance, the CDF command returns $F(z=1) = .8413$, which is the area to the left of this z=1 value.

INVCDF: The INVCDF command is the inverse of the CDF command: for a given or stated area (or probability) such as $F(z)=.50$ in the left-most tail, the INVCDF command returns the corresponding Z, which for this case is $z = 1$.

GRAPHS: the graphs shown below were generated from Minitab as described earlier, and then edited with the graph editor.   The lines, arrows, and hatching were all added to the graphs.

PROBLEM 8.3

a. $P(0 \leq Z \leq 1.08)$: type the commands

```
MTB > cdf 1.08;          #asking for the area to the left of z = 1.08
SUBC> normal mu 0 stdev 1.          #include the period
```

b.   The display is
```
Cumulative Distribution Function
Normal with mean = 0 and standard deviation = 1
      x        P( X <= x)
   1.0800         0.8599
```
c.   so the area to the left of z = 1.08 is .8599, and the desired area is
     .8599 - .5000 = .3599 .

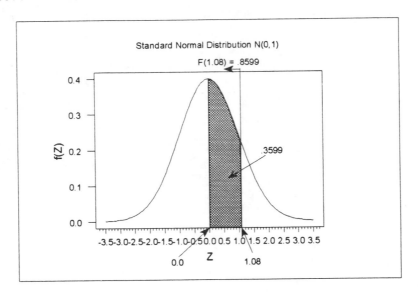

d.   P(Z < 0 or Z > 1 .08): the result from the first problem is used here:
     this is just 1 - .3599 = .6401 .

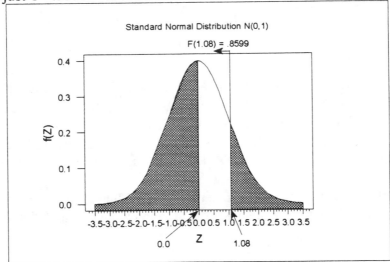

e.   To determine the probability that Z is between -.21 and the mean, we need the cdf for
     z = -.21. Type the commands in the Session Window
```
MTB > cdf -.21;
SUBC> normal mu 0 stdev 1.
```

f.   The resulting display is
```
Cumulative Distribution Function
Normal with mean = 0 and standard deviation = 1
    x        P( X <= x)
 -0.2100          0.4168
```

g.   The computations required are F(0) - F(-.21) = .5000 - .4168 = .0832 .

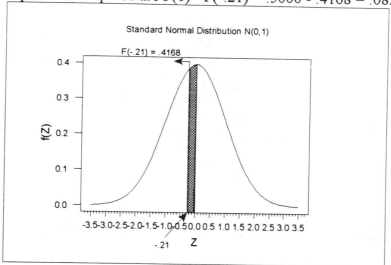

h.   To determine the probability that Z is less than -.21 or Z is greater than the mean (that is, Z > 0), we can use what we already have: .0832 was computed previously, and the desired probability is just 1 - .0832 = .9168 .

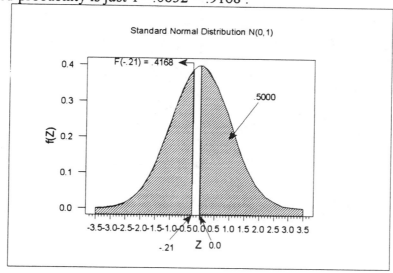

i.   For P(Z ≤ 1.08), this is just F(1.08), which was computed in the first problem to be .8599 .

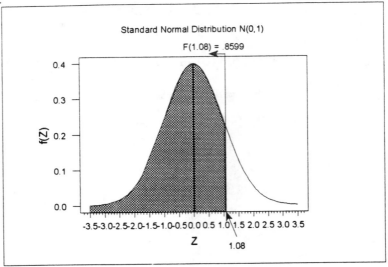

j.   To compute P(Z ≥ -0.21), this is just 1 - F(-.21) = 1 - .4168 = .5832 .

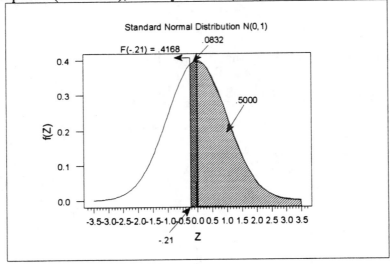

k.   For P(-0.21 ≤ Z ≤ 1.08), we have the required values available:
      F(1.08) - F(-0.21) = .8599 - .4168 = .4431 .

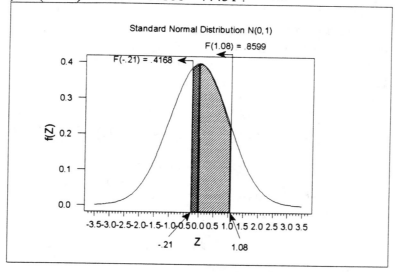

l.   For P(Z < -0.21 or Z > 1.08), we can break this up: [1] P(Z < -0.21) = F(-0.21) = .4168, and [2] P(Z > 1.08) = 1 - F(1.08) = .1401 . Therefore, P(Z < -.21 or Z > 1.08) = .4168 + .1401 = 0.5569 .

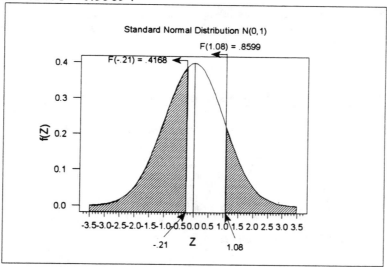

PART (b):

(1).   P(Z > 1.08) has already been computed: 1 - F(1.08) = 1 - .8539 = .1461 .

(2).   P(Z < -0.21) has already been computed: F(-0.21) = .4168 .

(3). $P(-1.96 \leq Z \leq -0.21)$: we have to get the CDF for $z = -1.96$.

    a.   Type in the Session Window the following commands

```
MTB > cdf -1.96;
SUBC> normal mu 0 stdev 1.
```

    b.   which gives this display, so $F(-1.96) = .0250$.

```
Cumulative Distribution Function
Normal with mean = 0 and standard deviation = 1
     x            P( X <= x)
  -1.9600         0.0250
```

    c.   Then $P(-1.96 \leq Z \leq -0.21) = F(-0.21) - F(-1.96) = .4168 - .0250 = .3918$.

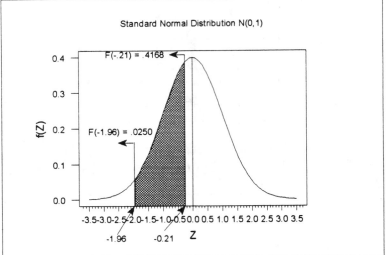

(4). $P(-1.96 \leq Z \leq 1.08)$: we have the needed results: $F(1.08) - F(-1.96) = .8599 - .025 = .8349$.

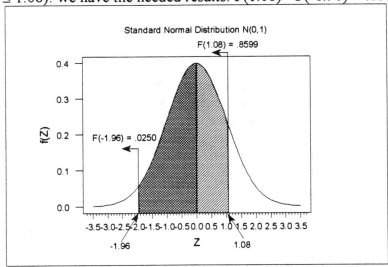

(5). P(1.08 ≤ Z ≤ 1.96): we need to compute F(1.96) - F(1.08). We have to get the CDF for
z = 1.96.
   a.   Type in the Session Window the following commands
        MTB > **cdf 1.96;**
        SUBC> **normal mu 0 stdev 1.**
   b.   which gives this display, so F(1.96) = .9750 .
        Cumulative Distribution Function
        Normal with mean = 0 and standard deviation = 1
             x   P(X <= x)
          1.9600   0.9750
   c.   Therefore, F(1.96) - F(1.08) = .9750 - .8599 = .1151 .

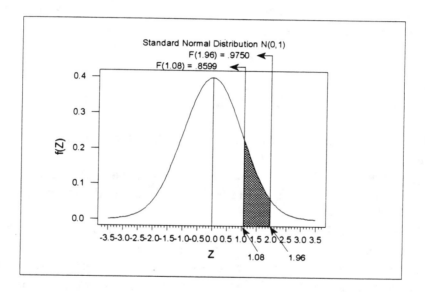

PART (c): What is the Z if 50% of the Z values are smaller than that Z? This is clearly z = 0,
   since at the center of the distribution (where z = 0) is where 50% of the values are smaller
   and 50% of the values are larger.

PART (d): What is the Z if 15.87% of the values are smaller than that Z? Here we need the
   INVCDF for the normal distribution which provides for a given probability in the left-most
   tail the corresponding Z value.
   a.   In the Session Window type the commands shown below.   (For this left-most tail
        area, what is the Z?)
        MTB > **invcdf .1587;**
        SUBC> **normal mu 0 stdev 1.**   #Include the period
   b.   The display is:
        Inverse Cumulative Distribution Function
        Normal with mean = 0 and standard deviation = 1
        P(X <= x)   x
           0.1587   -0.9998
   c.   Therefore, the Z for which 15.87% of the Z values are smaller is z = -.9998, or
        z = -1 .

PART (e): What is the Z if 15.87% of the values are larger than that Z? Since the results pertain to the left-most tail, the question must be transformed to this: What is the Z if 84.13% (100% - 15.87% = 84.13%) of the values are smaller than that Z? We need the INVCDF of .8413, so in the Session Window type the commands

```
MTB > invcdf .8413 ;
SUBC> normal mu 0 stdev 1 .
```

a.   The display is
```
Inverse Cumulative Distribution Function
Normal with mean = 0 and standard deviation = 1.00000
P( X <= x)           x
   0.8413          .9998
```

b.   Therefore, the Z for which 15.87% of the Z values are larger is the Z for which 84.13% of the Z values are smaller, which is z =.9998, or z = 1 .

PROBLEM 8.6

Food expenditures are normally distributed with $\mu_x = 420$ and $\sigma_x = 80$.

1.   Setup

a.   Setup the problem in Minitab. The range of data for our problem is about $\mu_x \pm 3\sigma_x$, which is $420 \pm 3(80)$, or (180 , 660).

b.   Place the heading *Z* in some column and *f(Z)* in the next column. To set the data of interest for this problem, type in the Session Window the following commands:
```
MTB > set 'Z'
DATA> 180:660        #generates numbers from 180 to 660 by 1's
DATA> end.           #include the period
```

c.   Place the corresponding normal values into the *f(Z)* column by typing the following commands in the Session Window, which will store the normal densities in *f(Z)*.
```
MTB > pdf 'z' 'f(z)';
SUBC> normal mu 420 stdev 80.
```

2.   Generate a plot

a.   Select **Graph>Plot**

b.   For Graph Variables: for Y double-click on *f(Z)*; for X double-click on *Z*.

c.   For Display, choose for Item 1 *Connect*, and for *For each*, choose *Graph*.

d.   Title: Click on **Annotation>Title** and type **Food Expenditures (Ex 8.6)**

e.   Click in the first box under *Text Size* and type **1** to reduce the title size. Click *OK*.

f.   Axis tick marks: Click on **Frame>Tick** and you will see a screen with the following headings. The numbers enclosed in a box are the ones to enter. In Row 1, enter **10** for the Number of Major ticks and enter **4** for the number of Minor Ticks. Then Click OK.

	Direction	Side	Positions	Number of Major	Number of Minor
1	X			10	4

g.   Click on *OK* twice to obtain the graph.

3.   PART (a): What % of the expenditures are less than $350? Or, what is the area in the left-most tail to the left of $350? Find F(350): in the session window type the following:

```
MTB > cdf 350;
SUBC> normal mu 420 stdev 80.
```

a.   The display is below, so that F(350) = .1908. (Note that this is an interpolated result, midway between two table probabilities. If you were using tables from the text, you would likely pick one of the tabled value rather than interpolating.)

```
Cumulative Distribution Function
Normal with mean = 420.000 and standard deviation = 80.0000
      x              P( X <= x)
   350.0000           0.1908
```

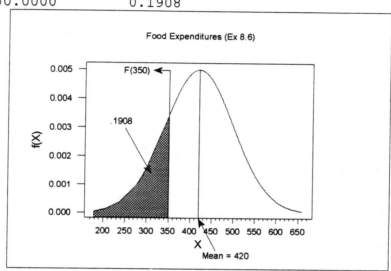

4.   PART (b): What % of the expenditures are between $250 and $350? Compute F(350) - F(250). Find F(250): in the session window type the following:

```
MTB > cdf 250;
SUBC> normal mu 420 stdev 80.
```

a.   The display is below, so that F(350) = .1908. (Note that this is an interpolated result, midway between two table probabilities.)

```
Cumulative Distribution Function
Normal with mean = 420.000 and standard deviation = 80.0000
      x              P( X <= x)
   250.0000           0.0168
```

b.   Therefore, F(350) - F(250) = .1908 - .0168 = .174: 17.4% of the expenditures are between $250 and $350.

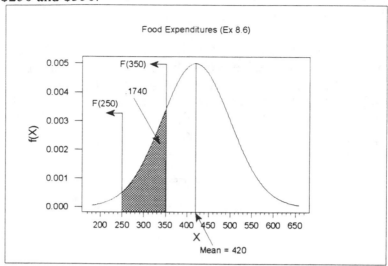

5.   PART (c): What % of the expenditures are between $250 and $450? Compute F(450) - F(250). Find F(450): in the Session Window type the following:

```
MTB > cdf 450;
SUBC> normal mu 420 sigma 80.
```

a.   The display is below, so that F(450) = .6462 .
```
Cumulative Distribution Function
Normal with mean = 420.000 and standard deviation = 80.0000
      x            P( X <= x)
   450.0000         0.6462
```
b.   Therefore, F(450) - F(250) = .6462 - .0168 = .6294, or 62.94% of the expenditures are between $250 and $350.

6.   PART (d): What % of the expenditures are less than $250 or greater than $450?

a.   Compute F(250) + (1 - F(450)) = .0168 + (1 - .6462 = .3538) = .3706 , or 37.06%.

7.   PART (e): Compute Q1 and Q3. Q1 is the 25th percentile, so use the INVCDF function to find what value of X has 25% of the values below it. Type in the Session Window:

```
MTB > invcdf .25;
SUBC> normal mu 420 sigma 80.
```

a.   The display is below, so that Q1 is 366.04. (Note that the associated Z value is z = .6745, which is interpolated, between two table values.)
```
Inverse Cumulative Distribution Function
Normal with mean = 420.000 and standard deviation = 80.0000
   P( X <= x)            x
     0.2500         366.0408
```
b.   Similarly, Q3 is the 75th percentile, so use the INVCDF function to find what value of X has 75% of the values below it. Type in the Session Window:

```
MTB > invcdf .75;
SUBC> normal mu 420 sigma 80.
```

c.   The display is below, so that Q3 is 473.96.
```
Inverse Cumulative Distribution Function
Normal with mean = 420.000 and standard deviation = 80.0000
   P( X <= x)              x
     0.7500           473.9592
```

ASSESSING THE NORMALITY ASSUMPTION AND CONSTRUCTION OF A NORMAL PROBABILITY PLOT

PROBLEM 8.26
Refer back to problems 3.6, 3.15, 3.29, and 3.37, which provide various results for the data in the file *cancer.dat*. In this section we'll extend the results which were provided there, in addition to generating a normal probability plot.

(Note that this data set (*Cancer*) is considered a *population* rather than a sample from a population. *Minitab* treats this population data as if it were sample data, so that for the standard .deviation it uses 49 (n-1=50-1=49) as the divisor rather than the correct 50. Even with this limitation, it is still easier to allow *Minitab* to analyze the data and then to do a separate computation for the standard deviation to obtain the correct population value.)

a.   The histogram in problem 3.15 for the *Cancer* data indicates the data appear to be normally distributed.
b.   Also in problem 3.15 the descriptive statistics were obtained. They are reproduced below (although the *StDev* value must be computed):

Variable	N	Mean	Median	TrMean	StDev	SEMean
Cancer	50	389.98	397.00	393.07	56.83	8.04

Variable	Min	Max	Q1	Q3
Cancer	229.00	500.00	363.00	435.00

c.   The Standard Deviation is shown as *StDev* = 56.83 = S. *Recall that this is the standard deviation if the data set were a sample*. Since it is a population, we can easily obtain the correct *StDev* in several ways.
   1.   There are several formulas for computing the variance and standard deviation for a population, and these could easily be used at the Session Window in *Minitab*.

2. One formula is $\sigma^2 = \dfrac{\Sigma(X^2) - \dfrac{(\Sigma X)^2}{N}}{N}$, which may be computed in *Minitab* using the *SSQ* function for computing the sum of squared values such as $\Sigma(X^2)$, and using *SUM* for computing the sum ΣX. Raising to a power in *Minitab* is accomplished using "**", such as **2 to square a value. Taking the square root employs the *SQRT* function. In the computation below we assign the population variance to the variable k1, and the standard deviation to the variable k2, and then both k1 and k2 are displayed using the *print* command.

```
MTB > let k1 = (SSQ(c4)-(SUM(c4)**2)/50)/50
MTB > let k2=SQRT(k1)
MTB > print k1-k2
```

a. The resulting display is shown below.

```
Data Display
K1        3165.18
K2        56.2599
```

3. Therefore the variance of the population is 3,165.18, and the standard deviation of the population is 56.26. This is the value which will be used to answer some questions which follow.

d. MODE: to determine the mode (the value which occurs the most), at the Session Window type

```
MTB > tally 'cancer'
```

e. and the list of values and their counts is given. There are 7 numbers which appear twice, so there is no single mode.

f. The Mean is 389.98, the median is 397.00, the midrange is (500 + 299)/2 = 399.50, the midhinge is (435 + 363)/2 = 399, and there is no single mode. All 5 are estimates of the central tendency. Note that there is agreement among the five estimators.

g. The *corrected population standard deviation* is 56.26, and the *Interquartile Range* is (435 - 363) = 72. Comparing 1.33 times the standard deviation, 1.33(56.26) = 74.83, with the IQ Range of 72, these measures are fairly close.

h. The range is (500 - 229) = 271, and six times the *corrected population standard deviation* is 6*56.26 = 337.56, so these values are not very close.

i. Next, we can look at the distribution of the data. Here we compute an interval and count the number inside the interval. One way is to sort the *Cancer* data into a new column and just count.

j. How many are within the range 389.98 ± 1(56.26) = (333.72 , 446.24)? There are 38/50 = 76% of the values within 1 *corrected population standard deviation*, which is close to the theoretical 67%.

k. How many are within the range 389.98 ± 1.28(56.26) = (317.97 , 461.99)? There are 41/50 = 82% of the values within 1.28 standard deviation, which is fairly close to the theoretical 80%.

l. How many are within the range 389.98 ± 2(56.26) = (277.46 , 502.50)? There are 48/50 = 96% of the values within 2 standard deviations, which is close to the theoretical 95%.

Normal Probability Plot

There are several methods for generating a normal probability plot.

1. One method is to generate normal scores and use a *Plot* command. For the data in the *Cancer* column, we need to generate normal scores and place them in some column, say c2, and provide a heading for c2 called *NormScor*.
    a. At the Session Window type

        ```
        MTB > nscores 'Cancer' c2      #gets normal scores, places in c2
        MTB > name c2 'NormScor'       #types column heading
        MTB > gStd.                    #prepares for a character graph
        MTB > plot 'Hours' 'NormScor'  #obtains the plot
        ```
    b. The Normal Probability Plot is shown below. This is close to a straight line, so the distribution appears to be close to normal.
        ```
        Character Plot
        ```

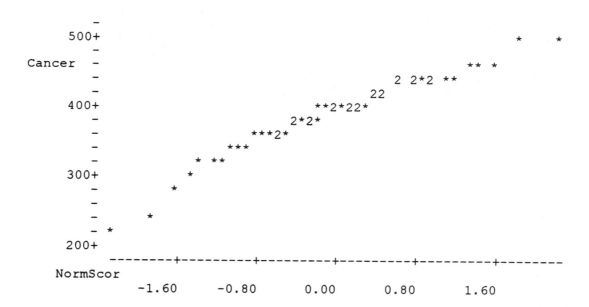

    c. Finally, type the following to restore *Professional Graphics*.
        ```
        MTB > gpro.
        ```
2. Another method for generating a normal probability plot entails the use of the %NormPlot macro or command. This method also provides a test for normality (Anderson-Darling test, or optionally Shapiro-Wilk test). (This approach is somewhat different from that in the text, and it is presented only for comprehensiveness.)
    a. Select **Graph>Normal Plot...**
    b. For *Variable* double-click on *Cancer*.
    c. Ignore the other options in the dialog box, including the title, since a suitable title will be generated.
    d. Click *OK.*

e. The display is shown below. (Note that the graph has been edited: the original value for *StdDev* in the lower left of the graph was 56.8311, which would be the standard deviation if this data set pertained to a sample. Since the *Cancer* data set pertains to a population, the *StdDev* value has been replaced by the correct *population standard deviation* of 56.26, and the word *Population* has been added to highlight this modification. Of course in this case the two standard deviations are essentially equal.)

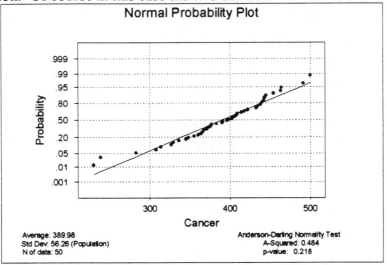

f. If you typed rather than using the menus,
    MTB > %NormPlot 'Cancer'.

SAMPLING DISTRIBUTIONS

This chapter develops the concept of a sampling distribution, and examines the important central limit theorem.

The methods for using sampling distributions for analysis are the same as for non-sampling distributions. Students frequently have difficulty distinguishing the two cases, but the difference can be stated as follows:

(1) In one case you are working with individual members of the population distribution, and it is individual values (X's) which are along the X-axis of the distribution. Questions pertain to individual values (the X's) and are of the form $P(X \geq x)$, and the standard deviation to use is the population standard deviation σ_x.

(2) A sampling distribution, on the other hand, pertains to *samples* of values from the population, and it is the *means* of the various possible samples ($\overline{X}$'s) which are along the X-axis of the distribution. Questions pertain to sample means and are of the form $P(\overline{X} \geq x)$, and the standard deviation to use is the population standard deviation σ_x

divided by the square root of the sample size, or $\sigma_{\overline{x}} = \dfrac{\sigma_x}{\sqrt{n}}$.

From the viewpoint of Minitab, there is no difference, since the same commands work for both. A typical set of commands is: `cdf 11; normal mu 8 sigma 2.` and from Minitab's viewpoint, it doesn't matter whether the `sigma 2` is σ_x or `sigma 2` is $\sigma_{\overline{x}}$.

SAMPLING DISTRIBUTION OF THE MEAN

PROBLEM 9.7
Long-distance calls are normally distributed with a population mean of $\mu_x = 8$ and a population standard deviation of $\sigma_x = 2$. Random samples are n = 25.

1.   Compute the standard deviation of the sampling distribution $\sigma_{\bar{x}} = \dfrac{\sigma_x}{\sqrt{n}} = \dfrac{2}{5} = .40$.

2.   What proportion of the sample means are between 7.8 and 8.2 minutes? Similar to the previous chapter, compute $P(7.8 \le \overline{X} \le 8.2) = F(8.2) - F(7.8)$.
   a.   Type at the session window:
```
MTB > cdf 8.2;
SUBC> normal mu 8 stdev .4.
```
   b.   The display is shown, where F(8.2) = 0.6915 .
```
Cumulative Distribution Function
Normal with mean = 8.00000 and standard deviation = 0.400000
      x           P( X <= x)
    8.2000          0.6915
```
   c.   Similarly, type
```
MTB > cdf 7.8;
SUBC> normal mu 8 stdev .4.
```
   d.   The display is shown, where F(7.8) = 0.3085 .
```
Cumulative Distribution Function
Normal with mean = 8.00000 and standard deviation = 0.400000
      x           P( X <= x)
    7.8000          0.3085
```
   e.   Therefore, $P(7.8 \le \bar{x} \le 8.2) = F(8.2) - F(7.8) = 0.6915 - 0.3085 = 0.3830$.

3.   What proportion are between 7.5 and 8 minutes? Compute $P(7.5 \le \overline{X} \le 8) = F(8) - F(7.5)$. Since the mean is 8, F(8) = .50. To find F(7.5),
   a.   at the Session Window type
```
MTB > cdf 7.5;
SUBC> normal mu 8 stdev .4.
```
   b.   The display is shown, where F(7.5) = .1056 .
```
Cumulative Distribution Function
Normal with mean = 8.00000 and standard deviation = 0.400000
      x           P( X <= x)
    7.5000          0.1056
```
   c.   Therefore, $P(7.5 \le \bar{x} \le 8) = F(8) - F(7.5) = 0.5000 - 0.1056 = 0.3944$.

4.   Use n = 100 and compute $P(7.8 \le \overline{X} \le 8.2)$: The new $\sigma_{\bar{x}} = \dfrac{\sigma_x}{\sqrt{n}} = \dfrac{2}{10} = 0.20$.

   Now compute $P(7.8 \le \overline{X} \le 8.2) = F(8.2) - F(7.8)$.
   a.   Type at the session window:
```
MTB > cdf 8.2;
SUBC> normal mu 8 stdev .2.
```

b.   The display is shown, where $F(8.2) = 0.8413$.
```
Cumulative Distribution Function
Normal with mean = 8.00000 and standard deviation = 0.200000
      x           P( X <= x)
    8.2000          0.8413
```

c.   Similarly, type
```
MTB > cdf 7.8;
SUBC> normal mu 8 stdev .2.
```

d.   The display is shown, where $F(7.8) = 0.1587$. Therefore,
$$P(7.8 \leq \overline{X} \leq 8.2) = F(8.2) - F(7.8) = 0.8413 - 0.1587 = 0.6826$$
```
Cumulative Distribution Function
Normal with mean = 8.00000 and standard deviation = 0.200000
      x           P( X <= x)
    7.8000          0.1587
```

5.   Shown below are plots of the two sampling distributions. The first is for the normal with mean 8 and standard deviation .4 (where n was 25) and the second is for the normal with mean 8 and standard deviation .2 (where n was 100). The difference in the two is obvious.

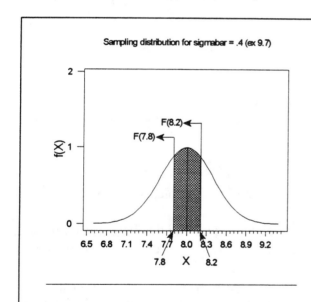 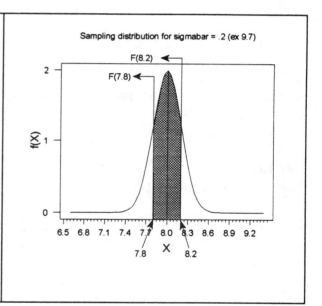

6.   Which is more likely?

(1) $P(X > 11)$, $\mu_x = 8$, $\sigma_x = 2$

(2) $P(\overline{X} > 9)$, $\mu_x = 8$, $\sigma_{\overline{x}} = \dfrac{2}{5} = 0.4$ with n = 25, or

(3) $P(\overline{X} > 8.6)$, $\mu_x = 8$, $\sigma_{\overline{x}} = \dfrac{2}{10} = 0.2$ with n = 100 ?

a.   To find $P(X>11) = 1 - F(11)$, type at the Session Window
```
MTB > cdf 11;
SUBC> normal mu 8 sigma 2.
```

b.   The display shows $P(X>11) = 1 - F(11) = 1 - 0.9332 = 0.0668$.
```
Cumulative Distribution Function
Normal with mean = 8.00000 and standard deviation = 2.00000
       x            P( X <= x)
    11.0000          0.9332
```
c.   To find $P(\overline{X} > 9) = 1 - F(9)$, type at the Session Window
```
MTB > cdf 9;
SUBC> normal mu 8 sigmabar .4.
```
d.   The display shows $P(\overline{X} > 9) = 1 - F(9) = 1 - 0.9938 = 0.0062$.
```
Cumulative Distribution Function
Normal with mean = 8.00000 and standard deviation = 0.400000
        x           P( X <= x)
     9.0000          0.9938
```
e.   To find $P(\overline{X} > 8.6) = 1 - F(8.6)$, type at the Session Window
```
MTB > cdf 8.6;
SUBC> normal mu 8 sigmabar .2.
```
f.   The display shows $P(\overline{X} > 8.6) = 1 - F(8.6) = 1 - 0.9987 = 0.0013$.
```
Cumulative Distribution Function
Normal with mean = 8.00000 and standard deviation = 0.200000
        x           P( X <= x)
     8.6000          0.9987
```
g.   Therefore, the most likely is $P(X>11) = 0.0668$.

SAMPLING DISTRIBUTION OF THE PROPORTION

The methods in Minitab for proportion problems are the same as for problems regarding means or individual values.

PROBLEM 9.13

Suppose that 93% of the deliveries arrive on time. Random samples of 500 deliveries are selected.

1.   What proportion of the time will the percentage of on-time deliveries be between 93% and 95%? (That is, how likely is it that the percentage of on-time delivery will be between 93% and 95%?)

a.   Notation: let the average proportion of on-time delivery be p = .93. The proportion is assumed to be normally distributed with a mean p = .93 and a standard deviation $\sigma_p =$

$\sqrt{\dfrac{p(1-p)}{n}} = \sqrt{\dfrac{.93(1-.93)}{500}}$. The sample proportion, or assumed sample proportion,

is called p_s. Finally, the Z value may be computed by $Z = \dfrac{p_s - p}{\sigma_p}$.

b.   Let Minitab compute these values: store σ_p in the constant K1, and store Z in the constant K2. Type the commands:
```
MTB > let k1 = sqrt(.93*.07/500)      #compute standard deviation
MTB > let k2 = (.95-.93)/k1           #compute the z value
MTB > print k1 k2
```

c.   The display gives the values.
```
Data Display
K1          0.0114105        #standard deviation
K2          1.75277          #z-value
```
d.   Therefore $P(.93 \le p_s \le .95) = F(.95) - F(.93) = F(.95) - 0.50$ (since .93 is the mean and $F(mean) = 0.50$).

e.   Type the commands (using the computed constant K1 as σ_p):
```
MTB > cdf .95;
SUBC> normal p .93 stdev k1.
```
f.   The display gives the values. Therefore, $P(.93 \le p_s \le .95) = F(.95) - F(.93) = F(.95) - 0.50 = 0.9602 - 0.5000 = 0.4602$. (Note that using a table would give a slightly different answer from the exact value just computed.)
```
Cumulative Distribution Function
Normal with mean = 0.930000 and standard deviation = 0.0114105
      x          P( X <= x)
   0.9500          0.9602
```
g.   To obtain a plot, first generate some data values. The data range is approximately $.93 \pm 3(.0114105) =$ which is approximately $(.89, .97)$. Suppose column c2 is available. Type the following
```
MTB > name c2 'p'  c3 'f(p)'      #types column headings
MTB > set 'p'                     #Set the horizontal axis values
DATA> .89:.97/.001
DATA> end
MTB > pdf 'p' 'f(p)';             #Set the density values
SUBC> normal p .93 sig k1.
MTB > print k1
```

Plot

To generate the plot using the menus:
a.   Select **Graph>Plot**
b.   For Graph Variables: for *Y* double-click on *f(p)*; for *X* double-click on *p*.
c.   For Display, choose for Item 1 *Connect*, and for *For each*, choose *Graph*.
d.   Title: Click on **Annotation>Title** and type **Sampling Distribution of Proportion (Ex 9.13)**
e.   Click in the first box under *Text Size* and type **1** to reduce the title size. Click *OK*.
f.   Axis tick marks: Click on **Frame>Tick** and you will see a screen with the following headings. The numbers enclosed in a box are the ones to enter. In Row 1, enter **10** for the Number of Major ticks and enter **9** for the number of Minor Ticks. Then Click OK.

	Direction	Side	Positions	Number of Major	Number of Minor
1	X			10	9

g.   Click on *OK* twice to obtain the graph. (Note that the graph has been edited.)

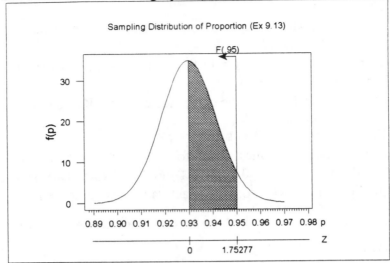

h.   If you typed rather than used the menus,
```
MTB >    Plot 'f(p)'*'p';
SUBC>    Connect;
SUBC>    Title "Sampling Distribution of Proportion (Ex 9.13)";
SUBC>     TSize 1;
SUBC>    Tick 1;
SUBC>      Number 10 9;
SUBC>    Tick 2.
```

i.   The answer to the next part of problem 9.13 is $P(p_s > .95) = 1 - F(.95) = 1 - 0.9602 =$ 0.0398 .

j.   Part 3 of 9.13 changes n to 1000. Letting Minitab do the computations:
```
MTB > let k1 = sqrt(.93*.07/1000)        #compute std dev
MTB > let k2 = (.95-.93)/k1              #compute Z value
MTB > print k1 k2
```

k.   The display is
```
Data Display
K1        0.00806846
K2        2.47879
```

l.   Compute the probability with the new standard deviation K1. Type the commands
```
MTB > cdf .95;
SUBC> normal p .93 stdev k1.
```

m.   The display below gives the values. Therefore, $P(.93 \le p_s \le .95) = F(.95) - F(.93) =$ F(.95) - 0.50 = 0.9934 - 0.5000 = 0.4934 . (Note that using a table would give a slightly different answer from the exact value just computed.)
```
Cumulative Distribution Function
Normal with mean = 0.930000 and standard deviation = 0.00806846
       x        P( X <= x)
    0.9500         0.9934
```

n.   The answer to the next part of problem 9.13 is $P(p_s > .95) = 1 - F(.95) = 1 - 0.9934 =$ 0.0066 .

CENTRAL LIMIT THEOREM
The general idea of the central limit theorem is this: if the sample size is large enough, the sampling distribution (the distribution of sample means, for instance) is normal, even if the population distribution from which the samples were drawn is not normal.

To demonstrate this, we'll suppose we have an exponential population distribution from which 1,000 samples of size 2 are taken. The mean is computed for each sample, and the 1,000 means comprise the sampling distribution. Then, take 1,000 samples of size 5, compute the mean of the 5 for each sample, and the 1,000 means again comprise the sampling distribution. Finally, do the same for 1,000 samples of size 30. Then, compare the plots of each such sampling distribution.

We should see that as the size of each sample increases, the sampling distribution of 1,000 samples looks more normally distributed about the population mean.
1.   To generate the data: Select **Calc>Random Data>Exponential** and in the *Generate rows* box type **1000**; in the *Store* box type **c1-c30**; leave the mean at 1.0 .
2.   To obtain the row means for different sample sizes:
    a.   get row means for samples of size 2 and place the 1,000 means in column 31: Select **Calc>Row Statistics** and click on *Mean*; click in the *Input Variables* box and type **c1-c2** (for samples of size 2); click in the *Store Result* box and type **c31**. Click *OK*.
    b.   get row means for samples of size 5 and place the 1,000 means in column 32: Select **Calc>Row Statistics** and click on *Mean*; click in the *Input Variables* box and type **c1-c5** (for samples of size 5); click in the *Store Result* box and type **c32**. Click *OK*.
    c.   get row means for samples of size 30 and place the 1,000 means in column 33: Select **Calc>Row Statistics** and click on *Mean*; click in the *Input Variables* box and type **c1-c30** (for samples of size 30); click in the *Store Result* box and type **c33**. Click *OK*.
3.   Provide headings for some of the columns, either by typing the heading directly above the column, or by typing at the Session Window the commands below. (We'll let column c30, now will be named *n=1* by the command below, pertain to the population, rather than to a sampling distribution.)
    a.   At the Session Window type:
```
MTB> name c30 'n=1' c31 'n=2' c32 'n=5' c33 'n=30'
```
4.   To generate the graphs for the *n=1* column
    a.   **Graph>Histogram**
    b.   For Graph Variables:, choose column headed *n=1*
    c.   For Display, choose *Symbol*, and for *For each*, choose *Graph*.
        Click on *Symbol* and click on the *Edit Attributes* button. For Type select *Solid Circle*. Make sure that the box under *Size* has the value 1.0. Click on *OK*.
    d.   Title: Click on **Annotation>Title** and type **Sampling Distribution from Exponential Population**
        Click in the box under the label *Text Size* and type **1** to reduce the title size. Click on *OK*.
    e.   Click on the Options button, and click on *Density* for *Type of Histogram*. Click on *Number of intervals* and click in the box and type **100**. Click on *OK*.

     f.   Click on **Frame>Min and Max:** click on the check box next to *X Minimum* then click in the box to its right and type **0**; for *X Maximum* type **4.5**; for *Y Minimum* type **0**; and for *Y Maximum* type **3**. Click on *OK*.

     g.   Click on *OK* to obtain the graph.

5.   Do the same for the *n=2*, *n=5*, and *n=30* columns, as well. (Note that rather than going through the menus for each case, you could go to the *Session Window* and copy the Minitab code which was generated from the previous graph and copy it to the newest MTB> prompt in the Session Window. You would have to change the heading from *n=1* to *n=2* or whatever the current desired column is. Then press Enter and the new graph would be generated.)

6.   The four graphs are shown below; they are all on the same scale in both the X and Y direction. Notice that the exponential *population* is not at all normal, but as the sample size of the 1,000 samples increases, the sampling distribution of sample means becomes more normal, and centered around 1, which is the mean of the original population exponential distribution.

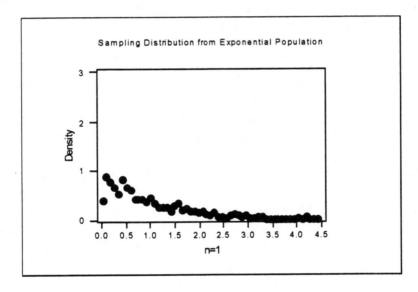

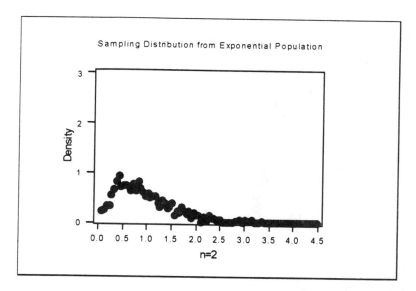

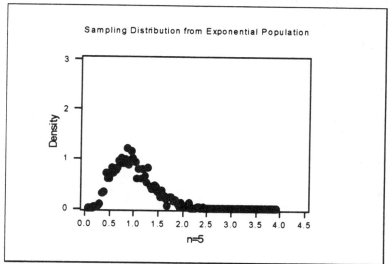

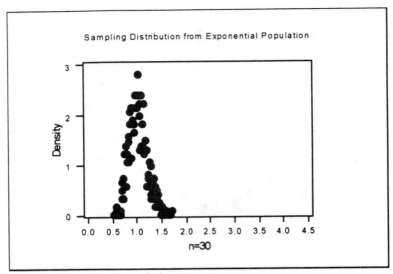

a. If you typed rather than used the menus,

```
MTB > Histogram 'n=1';
    Density;
    MidPoint;
    NInterval 100;
    Symbol;
      Type 6;
      Size 1;
    Title "Sampling Distribution from Exponential Population";
      TSize 1;
    Minimum 1 0;
    Maximum 1 4.5;
    Minimum 2 0;
    Maximum 2 3.
```

ESTIMATION

This chapter describes how to utilize the sampling distribution to develop confidence interval estimates for means and proportions.

CONFIDENCE INTERVALS
In order to take into consideration the variation inherent in estimating some parameter, the analyst frequently uses a confidence interval, which provides some specific statistical confidence in the estimate. The general form for most confidence intervals for estimating a parameter is:

sample estimate ± (some measure of confidence)*(some measure of variation).

CONFIDENCE INTERVAL FOR THE MEAN, USING Z
The formula for the confidence interval estimate for the unknown population mean is shown below

$$\overline{X} \pm Z \frac{\sigma_x}{\sqrt{n}}$$

PROBLEM 10.6
The goal is to estimate the average life of light bulbs. The standard deviation is known to be 100 hours. A random sample of n=50 bulbs yielded a sample average life of $\overline{x} = 350$ hours.
1.   Obtain a 95% confidence interval estimate of the true average amount. We assume the data are normally distributed.

a. To get the Z-value, note that we seek a 95% interval, so in Minitab find F(1-α/2) = F(.975). Type

```
MTB > invcdf .975;
SUBC> normal mu 0 stdev 1.
```

b. The display is as follows. The Z for 95% confidence interval is 1.96 .

```
Inverse Cumulative Distribution Function
Normal with mean = 0 and standard deviation = 1.00000
  P( X <= x)              x
    0.9750           1.9600
```

c. To employ Minitab to compute the confidence interval for the average light bulb life,

type the equations shown below. The confidence interval is $\overline{X} \pm Z\dfrac{\sigma_x}{\sqrt{n}}$. In the

equations below, $K1 = Z\dfrac{\sigma_x}{\sqrt{n}}$, K2 is the lower limit, and K3 is the upper limit. Type

```
MTB > let k1=1.96*100/sqrt(50)
MTB > let k2 = 350 - k1
MTB > let k3 = 350 + k1
MTB > print k1-k3
```

d. The result is shown below. The 95% confidence interval for the average light bulb life is (322.281 , 377.719).

```
Data Display
K1          27.7186
K2          322.281
K3          377.719
```

CONFIDENCE INTERVAL FOR THE MEAN, USING t

The t-distribution is very similar to the normal distribution. When the population standard deviation σ_x isn't known, it must be estimated using the sample standard deviation S. However, when the sample size is small, S may not be very accurate, and therefore the t-distribution is used, which accounts for more variation in the estimates. The population distribution must be normal, because the sample size when using t is frequently too small for the central limit theorem to apply. The confidence interval for the mean using the t distribution is

$$\overline{X} \pm t_{n-1}\frac{S}{\sqrt{n}} \ .$$

PROBLEM 10.19

The goal is to estimate the average life of flashlight batteries. The data are in the file Flashbat.dat. If the data aren't currently loaded, open the file by using the menu **File>Other Files>Import ASCII Data**; type **c1** for *column* and click *OK*. Give the file name *flashbat.dat* and its location.

1. Obtain a 95% confidence interval estimate of the true average amount. We assume the data are normally distributed.
   a. Select **Stat>Basic Statistics>1-Sample t...** and for *Variables* double-click on c1.
   b. Click on the *Confidence Interval* button and ensure the confidence *level* is set at 95 .

c.   The display is shown below. The 95% confidence interval for μ_x is (346.1, 600.9).

```
Confidence Intervals
Variable      N        Mean     StDev   SE Mean        95.0 % C.I.
C1           13       473.5     210.8      58.5  (   346.1,   600.9)
```

d.   If you typed rather than using the menus, at the Session Window type:
MTB > **TInterval 95.0 C1.**

BOOTSTRAP

The Bootstrap method of obtaining a confidence interval has the advantage of not requiring or imposing assumptions on the data or their distribution. The procedure is as follows: (1) obtain (or you are given) a sample of n values; (2) obtain a *new* sample of n values from the initial sample of n values, but this time *with replacement*, meaning there could be duplicates; (3) compute the mean of this new set of n values. Repeat steps (2) and (3) some k number of times, and collect the k means. These k means form their own distribution. Then, (4) sort this distribution of values; (5) the value at the smallest $\alpha/2*100\%$ position becomes the lower limit of the confidence interval, and the value at the largest $\alpha/2*100\%$ position becomes the upper limit of the confidence interval.

This method will require using a looping mechanism in Minitab. The best way of doing this is to create a macro using Minitab's macro language for these steps, but this is outside the scope of this book. Instead, we'll describe how to perform this looping using Minitab commands at the Session Window.

The general approach is to begin the looping process with the command *Store* (and provide a name for the set of commands being stored), which means *store the following commands for later execution*. Then type the commands to store, and then type *End*. Later, the command *Execute* (along with the name of the stored commands) will process the set of stored commands.

PROBLEM 10.26

These data represent a random sample of 13 values, taken without replacement from some population of lifetimes of flashlight batteries. We want to generate a set of 200 *resampled* means, where the steps (2) and (3) described above will be repeated 200 times, resulting in 200 means, which then form their own distribution. The result of typing at the Session Window the following commands will be a set of stored commands (named "sample") which may be executed as a block (essentially a macro). The column named "Means" will contain the 200 means after execution of the stored commands.

a.   Read in the *Flashbat.dat* data using (**File>Other Files>Import ASCII Data**; type **c1** for column and click *OK*. Or, if previously saved as *flashbat.mtw*, just open normally.

b.   Type the following at the Session Window.
```
MTB > names c1 'Initial' c2 'Resample' c3 'Means' c4
      'Sorted'
MTB > let k1=1
MTB > store 'sample'
STOR> sample 13 obvns from 'Initial' put 'Resample';
STOR> replace.
STOR> let 'Means'(k1)=mean('Resample')
STOR> let k1=k1+1
STOR> end
```

c.   Next, execute this block of commands 200 times. Type the following commands. (One could use the commands *NoEcho* to turn off the echo back to the screen, and then turn *Echo* back on at the end.)
```
MTB > let k1 = 1
MTB > execute 'sample' 200 times
```

d.   Now the column 'Means' contains the 200 means. Sort them and place them into the column *Sorted*.
```
MTB > sort 'Means' put 'Sorted'
```

e.   Since 200 samples (yielding 200 means) were generated, the lower confidence limit is the value at the smallest $\alpha/2*100\%$ of the values; the value at the $.025*100*200 = 5$th position, so it is the 5th smallest value, which for the means in the column 'Means' is 386.538. Similarly, the upper confidence limit is the 5 largest value, which is 588.462. Therefore, the 95% confidence interval for the unknown mean μ_x is (386.538 , 588.462).

PREDICTION INTERVAL FOR A FUTURE VALUE

An interval for predicting a particular future value is similar to the usual confidence interval. However, in a sense it is harder to predict a particular value than it is the mean of a distribution, so there is more variation in the possible answers, and therefore the standard deviation used in determining the prediction interval is larger than for a confidence interval. The prediction interval is shown below, where *t* is the statistic for the *Student's t-distribution*, and S is the sample standard deviation:

$$\overline{X} \pm t*S*\sqrt{1+\frac{1}{n}}\ .$$

PROBLEM 10.36

These data represent a random sample of 13 values, taken without replacement from some population of lifetimes of flashlight batteries.

a.   Read in the *Flashbat.dat* data using (**File>Other Files>Import ASCII Data**; type **c1** for column and *OK*. Or, if previously saved as *flashbat.mtw*, just open normally. Suppose the data are in column C1, with the heading **Hours.**

b.   Type the following at the Session Window.

```
MTB > name c1 'Hours'        #Or, just type the heading in at the column
MTB > describe 'Hours'       #To get the mean and standard deviation
```

    1.   Using the menus, Select **Stat>Basic Statistics>Descriptive Statistics** and double-click *Hours*.

c.   The display is shown below, where the mean is 473.5, the standard deviation is 210.8, and the sample size is 13.

```
Descriptive Statistics
Variable        N      Mean    Median    TrMean    StDev    SEMean
Hours          13     473.5     451.0     440.2    210.8      58.5

Variable      Min       Max        Q1        Q3
Hours       264.0    1049.0     307.5     553.5
```

d.   To get the t-value, note that we seek a 95% interval, so in Minitab find $F(1-\alpha/2) = F(.975)$, for $n-1 = 12$ degrees of freedom. Type

```
MTB > invcdf .975;
SUBC> t 12 df.
```

e.   The display is shown below. The correct t value is 2.1788.

```
Inverse Cumulative Distribution Function
Student's t distribution with 12 d.f.
P( X <= x)            x
   0.9750         2.1788
```

f.   To compute the prediction interval for a future battery life, type the equations shown below. The prediction interval is $\overline{X} \pm t * S * \sqrt{1+\dfrac{1}{n}}$. In the equations below,

$K1 = t * S * \sqrt{1+\dfrac{1}{n}}$, K2 is the lower limit, and K3 is the upper limit. Type the bolded portion shown below:

```
MTB > let k1=2.1788*210.8*sqrt(1+1/13)
MTB > let k2 = 473.5 - k1
MTB > let k3 = 473.5 + k1
MTB > print k1-k3
```

g.   The result is shown below. The lower limit is effectively 0. The 95% prediction interval for a future battery life is (0 , 950.129).

```
Data Display
K1       476.629
K2      -3.12888
K3       950.129
```

CONFIDENCE INTERVAL FOR A PROPORTION

Here we assume that for the binomial process being estimated, the normal approximation can apply. That is, assume both $np \geq 5$ and $n(1-p) \geq 5$. Then the confidence interval for the unknown population proportion p is shown below, where n is the number in the sample, p_s is the proportion in the sample, and Z is the normal variate (the Z value from the normal table):

$$p_s \pm Z^* \sqrt{\frac{p_s(1-p_s)}{n}} \ .$$

PROBLEM 10.45

The goal is to estimate the proportion of students with access to a computer outside the school. A sample of n=150 revealed that 105 had such access, so the sample proportion p_s = 105/150 = 0.70 . Obtain a 90% confidence interval for the true population proportion p.

    a.   To get the Z-value, note that we seek a 90% interval, so in Minitab find $F(1-\alpha/2)$ = F(.950). Type

```
MTB > invcdf .95;
SUBC> normal mu 0 stdev 1.
```

    b.   The display is shown below. The correct Z value is 1.6449 .

```
Inverse Cumulative Distribution Function
Normal with mean = 0 and standard deviation = 1
P( X <= x)          x
   0.9500        1.6449
```

    c.   To compute the confidence interval for the proportion with computer access, type the equations shown below. The confidence interval is $p_s \pm Z^* \sqrt{\frac{p_s(1-p_s)}{n}}$. In the equations below, $K1 = Z^* \sqrt{\frac{p_s(1-p_s)}{n}}$, K2 is the lower limit, and K3 is the upper limit. Type

```
MTB > let k1=1.6449*sqrt(.7*.3/150)
MTB > let k2 =  .7 - k1
MTB > let k3 =  .7 + k1
MTB > print k1-k3
```

    d.   The result is shown below. The 90% prediction interval for the true population proportion of students with computer access outside the school is (.638 , .762) .

```
Data Display
K1         0.0615465
K2         0.638453
K3         0.761546
```

SAMPLE SIZE FOR THE MEAN

To determine the minimum sample size required to achieve a specified statistical confidence in the estimate, the formula is the familiar Z formula except that Z is known and one solves for n. In order to achieve this confidence, the sample size if fractional must be rounded up. The allowable error is called *e*. The formula is then

$$n = \frac{Z^2 \sigma_x^2}{e^2}.$$

PROBLEM 10.50

The objective is to determine the minimum sample size required to provide an estimate of the average monthly electric bills, so that there would be a 99% confidence in the estimate. Based on other studies, the standard deviation is assumed to be \$25, and the desired allowable error in the estimate is ± \$5.

   a.   To get the Z-value, note that we seek a 99% interval, so in Minitab find $F(1-\alpha/2) = F(.995)$. Type
```
MTB > invcdf .995;
SUBC> normal mu 0 stdev 1.
```
   b.   The display is shown below. The correct Z value is 2.5758.
```
Inverse Cumulative Distribution Function
Normal with mean = 0 and standard deviation = 1
P( X <= x)          x
0.9950           2.5758
```
   c.   To compute the sample size, type the formula below, where k1 is $n = \frac{Z^2 \sigma_x^2}{e^2}$:
```
MTB > let k1 = 2.5758**2*25**2/(5**2)
MTB > print k1
```
   d.   The display is shown below. The required minimum sample size is 166 (rounding up). (Note that if the table value of 2.58 were used instead of 2.5758, the sample size would round up to 167.)
```
Data Display
K1           165.869
```

SAMPLE SIZE FOR A PROPORTION

The formula for computing the sample size for a proportion is similar to that for the sample size for the mean.

$$n = \frac{Z^2 p(1-p)}{e^2}.$$

PROBLEM 10.56

The objective is to determine the minimum sample size required to provide an estimate of the proportion of customers who would purchase a guide, so that there would be a 95% confidence in the estimate. Based on other studies, the proportion of those willing to purchase a guide is 0.30. The desired allowable error in the estimate is ± .05 of the true proportion.

a. To get the Z-value, note that we seek a 95% interval, so in Minitab find F(1-α/2) = F(.975). Type
```
MTB > invcdf .975;
SUBC> normal mu 0 stdev 1.
```

b. The display is shown below. The correct Z value is 1.96.
```
Inverse Cumulative Distribution Function
Normal with mean = 0 and standard deviation = 1
P( X <= x)          x
0.9750           1.9600
```

c. To compute the sample size, type the formula below, where k1 is $n = \dfrac{Z^2 p(1-p)}{e^2}$:
```
MTB > let k1 = 1.96**2*0.30*0.70/(0.05**2)
MTB > print k1
```

d. The display is shown below. The required minimum sample size is 323 (rounding up).
```
Data Display
K1          322.694
```

11
FUNDAMENTALS OF HYPOTHESIS TESTING

This chapter develops hypothesis-testing methodology as a technique for analyzing differences and making decisions, while evaluating statistical risks associated with decisions.

The objective in hypothesis testing is to test some inference. The general overview is: Make an assumption about a population parameter, perform an experiment to test it, do some calculations to see if the results are probable enough given the assumption. If not, throw out the assumption.

The assumption about the population parameter is stated in terms of the null hypothesis (H_0) of no difference, and the alternate hypothesis (H_1). These two hypotheses are stated in such a way that they cover the universe of possibilities.

One approach to stating the form of the two hypotheses which is frequently helpful to students is this: typically the statement of the problem provides one hypothesis, and it is then necessary to decide whether this hypothesis pertains to H_0 or H_1 . The rule of thumb is that the null hypothesis always has the equal sign.

In hypothesis testing, there are two types of error which may be made: A Type I error is *rejecting a true hypothesis* and a Type II error is *failing to reject a false hypothesis*. The associated probabilities are α = P(Making a Type I error) and β = P(Making a Type II error).

TWO-TAILED TEST

PROBLEM 11.10

Is the strength of the cloth 70 pounds or not? It is assumed that the population mean is 70 and the population standard deviation is 3.5 pounds. The sample of n=36 revealed a mean of $\bar{x}$ = 69.7 pounds. Test this at α = 0.05 level of significance.

    a.   The hypotheses are H_0: μ_x = 70 Versus H_1: $\mu_x \neq$ 70.

    b.   Since we have a summary of the data values rather than the actual values, Minitab cannot directly perform the hypothesis test; we have to give it formulas. Let Minitab determine the correct Z value, which is F(1-α/2) = F(.975). Type

```
MTB > invcdf .975;
SUBC> normal mu 0 stdev 1.
```

    c.   The display is shown below. The correct Z value is 1.96, which is the test value. (This could be looked up in a table.)

```
Inverse Cumulative Distribution Function
Normal with mean = 0 and standard deviation = 1
P( X <= x)          x
0.9750           1.9600
```

    d.   Now to decide whether the sample provides sufficient support to reject H_0, we compare the test Z-value with the sample Z-value. If the sample result converted to a Z-score is less than -1.96 or more than 1.96, this will be sufficient evidence to reject H_0 in favor H_1. Transform the sample result to a Z-value: type

```
MTB > let k1 = (69.7-70)/(3.5/sqrt(36))
MTB > print k1
```

    e.   The result is shown below. Since the sample Z value is not less than -1.96, then one cannot reject H_0, so the conclusion is that the strength of the cloth is probably 70 pounds. (Or, the evidence isn't strong enough to reject H_0.)

```
Data Display
K1          -0.514291
```

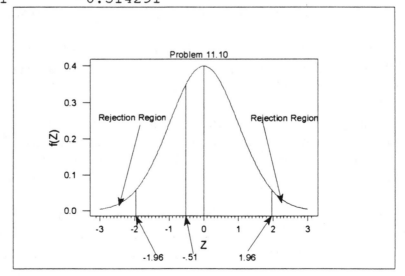

p-value

The p-value is useful in hypothesis testing, largely because many computer programs routinely provide it. In one sense it may provide the answer to a typical hypothesis-testing question. The p-value is the tail area for the sample Z-value, if using the normal distribution. (If it is a two-tailed test, the p-value is twice the tail area for the sample Z-value.) Using the procedures provided in this workbook, here is a suggested method for determining the correct p-value:

1.   For a left-tailed test,   p-value = CDF(X)
2.   For a right-tailed test,   p-value = 1 - CDF(X)
3.   For a two-tailed test,   p-value = 2 * minimum of { CDF(X) , 1 - CDF(X) }

PROBLEM 11.18

Referring to problem 11.10, compute the p-value. The p-value is the tail area from the sample Z value.

a.   For problem 11.10, the sample Z value (the sample $\bar{x}$ value converted to a Z) was -.51. The tail area for z = -0.51 is F(-0.51). Type
MTB > **cdf -.51;**
SUBC> **normal mu 0 stdev 1 .**

b.   The result is shown below. The tail area for the sample Z of -0.51 is 0.3050. Since this was a two-tailed test, the p-value is twice the tail area, so the p-value is 0.61 . Since this represents a high risk of a Type I error if the null hypothesis is rejected, we were right to not reject H_0 in the previous problem.

```
Cumulative Distribution Function
Normal with mean = 0 and standard deviation = 1.00000
        x          P( X <= x)
     -0.5100            0.3050
```

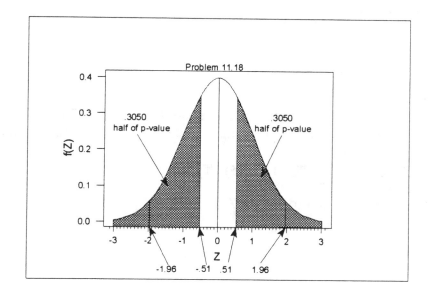

ONE-TAILED TEST

PROBLEM 11.29
Do the steel bars have an average length of *at least* 2.8 feet? It is assumed that the population mean is 2.8 feet and the population standard deviation is 0.20 feet. The sample of n = 25 revealed a mean of $\bar{x} = 2.73$ feet. Test this at $\alpha = .05$ level of significance.

a.　H_0: $\mu_x \geq 2.80$ versus H_1: $\mu_x < 2.80$.

b.　Let Minitab determine the correct Z value. The left-hand tail area with the Z value to the left of the mean is $F(\alpha) = F(0.05)$.
```
MTB > invcdf .05;
SUBC> normal mu 0 stdev 1.
```

c.　The display is shown below. The correct Z value is -1.6449, which is the test value.
```
Inverse Cumulative Distribution Function
Normal with mean = 0 and standard deviation = 1
P( X <= x)            x
0.0500          -1.6449
```

d.　Now to decide whether the sample provides sufficient support to reject H_0, we compare the test Z-value with the sample Z-value. If the sample result converted to a Z-score is less than -1.6449, this will be sufficient evidence to reject H_0 in favor H_1. Transform the sample result to a Z-value:
```
MTB > let k1 = (2.73-2.80)/(.20/sqrt(25))
MTB > print k1
```

e.　The result is shown below. Since the sample Z value of -1.75 is less than -1.6449, then one is lead by this sample result to reject H_0. So the conclusion is that the average length of steel bars is not at least 2.8 feet, and it is likely that the average length is less than 2.8 feet (since the sample result was less than that).
```
Data Display
K1          -1.7500
```

p-value

PROBLEM 11.36
Referring to problem 11.10, 11.18, and 11.29, compute the p-value. The p-value is the tail area from the sample Z value. For problem 11.29, the sample Z value (the sample $\bar{x}$ value converted to a Z) was -1.75.

a.　The tail area for z=-1.75 is F(-1.75). Type
```
MTB > cdf -1.75;
SUBC> normal mu 0 stdev 1 .
```

b.   The result is shown below. The tail area for the sample Z of -1.75 is 0.0401. Since this was a one-tailed test, this tail area is the p-value, so the p-value is 0.0401 . Because this represents a small risk of a Type I error if the null hypothesis is rejected, we were right to reject H_0 in the previous problem. The p-value is less than a typical α level of risk of 0.05. Since the risk of an error when rejecting H_0 is 0.0401 < 0.05, it is reasonable to reject H_0.

```
Cumulative Distribution Function
Normal with mean = 0 and standard deviation = 1.00000
    x        P( X <= x)
 -1.7500          0.0401
```

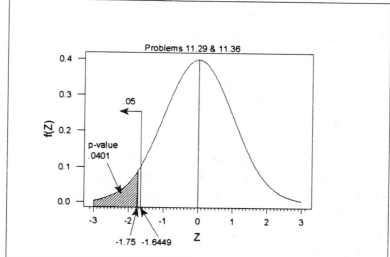

Using Minitab's ZTEST command and ZINTERVAL command

For problem 11.29 it was necessary to compute the Z value in a calculator-like fashion, since the ZTEST is pertinent only for raw data, not summary data. For this example we'll generate data with the same mean and standard deviation, in order to demonstrate both the ZTEST command and the method for generating normal data.

a.   First, generate 500 values which are normally distributed with a mean of 2.8 and a standard deviation of 0.20 (and store in column c5). Select **Calc>Random data>Normal** and in the dialog box which appears type the values in **bold** shown below. (Note that if you wished for the random numbers to be reproducible (to obtain the same numbers if you later performed the same steps), you would first select **Calc>Set Base...** and enter an integer such as 100.)

1.   Generate **500** rows of data
2.   Store in column **c5**
3.   Mean   **2.8**
4.   Standard deviation   **.20**
5.   Click on *OK*.

6.   If you typed rather than using the menus,
```
MTB > Random 500 c5;
SUBC>   Normal 2.8 .20000.
```

b.   The resulting 500 values are stored in Column c5. To verify that we have 500 values with a mean of 2.8 and a standard deviation of 0.20, we use the *Describe* command. The menu approach is to select **Stat>Basic Statistics>Descriptive Statistics**, and at the dialog box which appears double-click on the column c5 in the list, and click on *OK*.

       1.   If you typed rather than using the menus,
```
MTB > describe c5
```

       2.   The display is shown below. Note that the mean of the 500 values is 2.7933 and the standard deviation is 0.204 .
```
Descriptive Statistics
```

Variable	N	Mean	Median	TrMean	StDev	SEMean
C5	500	2.7933	2.7869	2.7939	0.2040	0.0091

Variable	Min	Max	Q1	Q3
C5	2.2058	3.3965	2.6565	2.9325

c.   Next, generate a sample of 25 values from this set of 500 values and store the 25 values in Column c6. Select **Calc>Random data>Sample From Columns** and in the dialog box which appears type the values in **bold** shown below.

       1.   Sample **25** rows from column

       2.   **c5**   (enter this in the next box)

       3.   Store samples in   **c6**

       4.   Click on *OK*.

       5.   If you typed rather than using the menus,
```
MTB > Sample 25   c5 c6.
```

d.   The resulting 25 values are stored in Column c6. To observe the statistics for these 25 values, we use the *Describe* command. The menu approach is to select **Stat>Basic Statistics>Descriptive Statistics**, and at the dialog box which appears double-click on the column c6 in the list, and click on *OK*.

       1.   If you typed rather than using the menus,
```
MTB > describe c6
```

       2.   The display is shown below. Note that the mean of the 25 values is 2.7687 and the standard deviation is 0.2037 . (Since the process we've pursued provides a random set of numbers, your results may differ slightly.)
```
Descriptive Statistics
```

Variable	N	Mean	Median	TrMean	StDev	SEMean
C6	25	2.7687	2.7516	2.7718	0.2037	0.0407

Variable	Min	Max	Q1	Q3
C6	2.2910	3.1761	2.6458	2.9266

e.   Finally, with the 25 sample values in Column c6 we wish to test the hypotheses as in Problem 11.29: H_0: $\mu_x \geq 2.80$ versus H_1: $\mu_x < 2.80$. In Problem 11.29 we performed the computations directly, since we only had summary values. Now that we have some data (which we generated), we can use the ZTEST command. Using the menus, select **Stat>Basic Statistics>1-Sample z**, and for the dialog box which appears follow the steps below.

1.   For variables double-click on `c6`
2.   Click on the *Mean* button and type   `2.8`
3.   For the *Alternative* click on the arrow and select `less than`
4.   For *Sigma* type   `.2`
5.   Click on *OK*.
6.   If you typed rather than using the menus,
```
MTB > ZTest 2.8 .2  C6;
SUBC>    Alternative -1.
```
7.   The display is shown below
```
Z-Test
Test of mu = 2.8000 vs mu < 2.8000
The assumed sigma = 0.200

Variable      N      Mean    StDev   SE Mean       Z    P-Value
C6           25    2.7687   0.2037    0.0400   -0.78       0.22
```

f.   If Problem 11.29 had provided the actual sample data rather than summary statistics, we would have used the *ZTEST* command as described above, rather than computing the Z statistic. The result here is similar to that in Problem 11.29, but of course it differs in that the sample mean of Problem 11.29 was stated as 2.73 where for our random sample it was 2.7687. The standard deviations were of course different, as well. The goal of this exercise was to indicate how to generate normal data, and once those data are available, to provide the steps for using the ZTEST command.

g.   Next we wish to construct a 95% confidence interval for the unknown population mean, using the 25 sample values in Column c6. Using the menus, select **Stat>Basic Statistics>1-Sample z**, and for the dialog box which appears follow the steps below.

1.   For *variables* double-click on `c6`
2.   Click on the *Confidence Interval* button
3.   Ensure the Confidence *Level* is set at **95** .
4.   For *Sigma* type   `.2`
5.   Click on *OK*.
6   If you typed rather than using the menus,
```
MTB > ZInterval 95.0 .2 C6.
```
7.   The display is shown below
```
Confidence Intervals
The assumed sigma = 0.200

Variable N    Mean    StDev  SE Mean       95.0 % C.I.
C6       25  2.8068   0.2038   0.0400   ( 2.7283,  2.8852)
```

TYPE II ERRORS AND POWER OF THE TEST

The probability of a Type I error is α, so that P(Rejecting true H_0) = α. One might wonder why not just make α very low, or zero, to avoid that risk. The answer is that lowering the α risk raises the other type of risk; the probability of a Type II error. The P(Type II: Failing to reject a false H_0) = β. One approach in statistics is to try to balance these two probabilities, but generally one error is more important to protect against than another, so frequently the α error is the one of interest.

In this section we'll look at this other type of risk, the probability of being led by the sample to make a wrong decision by not rejecting H_0 when it is false, which is a Type II error. The *power of the test*, conversely, is the probability of making the correct decision when H_0 is false; the correct decision is to reject this false H_0, and the probability of making this correct decision is called the power of the test, which is $1 - \beta$.

PROBLEM 11.42
A machine should fill at least 7 oz of beverage in a cup, with a standard deviation of 0.2 oz. A random sample of $n = 16$ cups is taken, and the acceptable risk level is $\alpha = 0.05$.

- a. H_0: $\mu_x \geq 7$ versus H_1: $\mu_x < 7$.
- b.   Let Minitab determine the correct Z value. The left-hand tail area with the Z value to the left of the mean is $F(\alpha) = F(.05)$.
  ```
  MTB > invcdf .05;
  SUBC> normal mu 0 stdev 1.
  ```
- c.   The display is shown below. The correct Z value is -1.6449, which is the test value.
  ```
  Inverse Cumulative Distribution Function
  Normal with mean = 0 and standard deviation = 1
  P( X <= x)            x
  0.0500            -1.6449
  ```
- d.   Test the hypothesis if the sample value is $\bar{x} = 6.9$ oz. First, convert the sample result 6.9 to a Z value by typing
  ```
  MTB > let k1 = (6.9 - 7.0)/(.20/sqrt(16))
  MTB > print k1
  ```
- e.   The result is shown below. The sample $\bar{x} = 6.9$ corresponds to a sample Z of -2.00 .
  ```
  Data Display
  K1          -2.0000
  ```
- f.   You've already set up the critical value for rejecting H_0: if $Z_{sample} < -1.6449$ you will reject H_0. Since $Z_{sample} = -2.00$, we would reject H_0. Keep this same critical value (-1.6944) for rejecting H_0 but now assume the true mean is not 7.0 but is instead 6.9. With the same critical value, now how likely is it that you will make a mistake by not rejecting H_0? (This probability is β.) Similarly, how likely is it, now that the true mean is assumed to be 6.9, that you will be lead by the sample result to make the correct decision: to reject this false H_0? (This probability is $1 - \beta$ and is called the power of the test - the likelihood of correctly rejecting a false H_0.)
- g.   The new distribution is centered at 6.9, which corresponds to a Z of -2.0 in the original graph (but the centerline of the *new* graph is Z=0, of course). Similarly, the original critical value of -1.6449 corresponds to a Z in this new graph of .3551. Therefore we can use Minitab to find, for the new distribution, the left tail for the Z-value of .3551. Type
  ```
  MTB > cdf .3551 ;
  SUBC> normal mu 0 stdev 1 .
  ```

h.   The result is shown below.
     ```
     Cumulative Distribution Function
     Normal with mean = 0 and standard deviation = 1.00000
             x         P( X <= x)
          0.3551          0.6387
     ```

i.   The conclusion is that if the true mean were 6.9 rather than 7.0, the probability of correctly rejecting a false H_0 is $1-\beta = 0.6387$, which is the power of this test. In addition, the probability of not rejecting a false H_0 is $\beta = 0.3613$.

j.   The following graph shows the overlaid curves with $\alpha = 0.05$ and $\beta = 0.6387$.

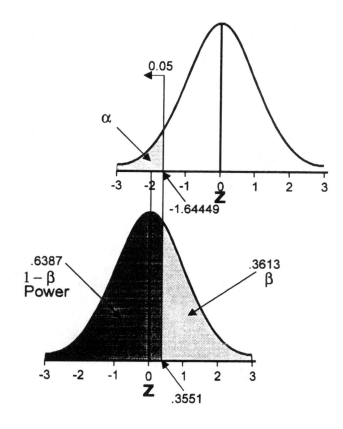

ONE-SAMPLE TESTS WITH NUMERICAL DATA

·The goal of this chapter is to extend the principles of hypothesis testing to the common one-sample test for the mean, median, and variance. In addition, nonparametric procedures are discussed.

ONE-SAMPLE t TEST

PROBLEM 12.2

Test at $\alpha=.05$ whether the average plastic hardness is greater than 260 units, using the n = 50 values in *Plastic.dat*. Use the menu **File>Other Files>Import ASCII Data**; type **c1** for *column* and click *OK*. Give the file name *Plastic.dat* and its location. Type the heading **X** for the column of data.

    a.   The hypotheses are H_0: $\mu_x \leq 260$ versus H_1: $\mu_x > 260$.

    b.   Since we have the actual data values rather than summary information, we can let Minitab do the computations for us.

    c.   Select the menu **Stat>Basic Statistics>1-Sample t**: for variables type or double-click *X*; click on *Test Mean* and click on its box and type **260**; For the Alternate (hypothesis) choose *Greater than* and click *OK*. The Minitab commands you would type if you didn't use the menus are:

```
MTB > TTest 260 'X';
SUBC>    Alternative 1.
```

d.   The Display is shown

```
T-Test of the Mean
Test of mu = 260.00 vs. mu > 260.00
Variable      N       Mean     StDev    SE Mean        T      P-Value
X            50     267.64     24.44       3.46     2.21        0.016
```

e.   Note that the t value is 2.21, and the p-value is 0.016. Without any more work we know to reject H_0 because the p-value of 0.016 is far less than any reasonable α such as 0.05 .

f.   (Optional: If we didn't use the Minitab *1-sample t* test, but instead gave Minitab formulas to compute, the following steps would be necessary: we would obtain the mean and standard deviation using MTB > **Describe 'X'**; transform the sample $\bar{x} = 267.64$ to a t-value by computing MTB > **let k1 = (267.64 - 260)/(24.44/sqrt(50))**, which would give a t-value of 2.21. We would find the critical t-value by typing MTB > **invcdf .95; SUBC> t 49 df** , and this would give the critical t-value of 1.67766. Then, the sample t-value of 2.21 > critical t-value of 1.67766, so reject H_0.)

WILCOXON SIGNED-RANKS TEST OF HYPOTHESIS FOR THE MEDIAN

This test is used for testing hypotheses about the median, rather than the mean as we have been doing. The only assumption is that the underlying distribution is symmetrical. When assumptions for using the t distribution are not met, the Wilcoxon Signed-Ranks test is an alternative.

PROBLEM 12.13

Is there evidence that the median amount in larceny claims has increased from \$125? A sample of 18 claims is taken. Test at α=.05.

a.   The hypotheses are H_0: median $\le$ 125 versus H_1: median > 125.

b.   Read the data from the data file using the menu: **File>Other Files>Import ASCII Data**; type **c1** for *column* and click *OK*. Give the file name *Claim.dat* and its location. Type the heading **Claims** for the column of data.

c.   First, for comparison, treat the problem as a t-distribution problem, and test for a difference in the mean. Select the menu **Stat>Basic Statistics>1-Sample t**: for variables type or double-click *Claims*; click on *Test Mean* and click on its box and type **125**; For the Alternate (hypothesis) choose *Greater than* and click *OK*.

d.   The Minitab commands you would type if you didn't use the menus are shown below.

```
MTB > TTest 125 'Claims';
SUBC>    Alternative 1.
```

e.   The display is shown below. Since the p-value of 0.16 is much larger than a typical α of .05 or so, we would conclude that there is insufficient evidence to reject H_0.

```
T-Test of the Mean
Test of mu = 125.00 vs mu > 125.00
Variable      N       Mean     StDev    SE Mean        T      P-Value
Claims       18      141.4      68.3       16.1     1.02         0.16
```

f. Now use the Wilcoxon test for the median. Select the menu
    Stat>Nonparametrics>1-Sample Wilcoxon: for variables type or double-click
    Claims; click on *Test median* and click on its box and type **125**; For the Alternate
    (hypothesis) choose *Greater than* and click *OK*.

g. The Minitab commands you would type if you didn't use the menus are:
```
MTB > WTest 125  'Claims';
SUBC>    Alternative 1.
```

h. The display is shown. The Wilcoxon statistic is 91.5, the sample median is 127.5, and
    the p-value is 0.405. With a p-value this large we would be led to not reject H_0.
    There is no evidence that the median size of claim has increased.
```
Wilcoxon Signed Rank Test
TEST OF MEDIAN = 125.0 VERSUS MEDIAN G.T. 125.0
 N FOR    WILCOXON                    ESTIMATED
            N   TEST  STATISTIC  P-VALUE    MEDIAN
Claims     18    18       91.5    0.405      127.5
```

χ^2 TEST OF HYPOTHESIS FOR THE VARIANCE

The statistic for testing the variability in the population is $\chi^2 = \dfrac{(n-1)S^2}{\sigma_x^2}$.

PROBLEM 12.24

Doorknobs tend to be 2.5 inches diameter, and the previous standard deviation has been 0.035
inch. A sample of 25 doorknobs was produced, with a 0.025 standard deviation. Is there
evidence that the population standard deviation is now less than 0.035, or that the population
variance is less than $(0.035)^2$?

a. The hypotheses are: H_0: $\sigma_x^2 \geq (0.035)^2$ versus H_1: $\sigma_x^2 < (0.035)^2$.

b. The commands in Minitab to compute the statistic are shown below.
```
MTB > let k1 = (25-1)*(.025**2)/(.035**2)
MTB > print k1
```

c. The result is shown below. The sample statistic $\chi^2 = 12.2440$.
```
Data Display
K1    12.2440
```

d. This is a one-tailed test in the left tail, so we need F(.05). To compute the critical χ^2
    value, type
```
MTB > invcdf .05;
SUBC> chisquare 24 df .
```

e. The display is shown below. Since the (sample $\chi^2 = 12.244$) < (critical $\chi^2 = 13.8484$),
    reject H_0. There is sufficient sample evidence that the standard deviation is less than
    0.035.
```
Inverse Cumulative Distribution Function
Chisquare with 24 d.f.
P( X <= x )         x
   0.0500      13.8484
```

One-Sample Tests: Numerical Data

p-value

a.   To compute the p-value for this test, we need the tail area for the sample $\chi^2 = 12.244$.
MTB > **cdf 12.244;**
SUBC> **chisquare 24 df .**

b.   The display is shown below. The tail area for $\chi^2 = 12.244$ is 0.0230, which is the tail area for the sample χ^2, which in this case is the p-value. Since the p-value is so small, this is evidence that H_0 should be rejected.
```
Cumulative Distribution Function
Chisquare with 24 d.f.
        x      P( X <= x)
   12.2440        0.0230
```

WALD-WOLFOWITZ ONE-SAMPLE RUNS TEST FOR RANDOMNESS

The hypotheses are H_0: the process is random, versus H_1: the process is not random. The approach for testing is to observe the order in which the items in the sample are obtained. Then, look for runs, where the same type item appears clustered or in some order, which would indicate a nonrandom process.

PROBLEM 12.34

For the 30 pairs of jeans, is there evidence the process is out of control in terms of the desired length of 34 inches?

a. H_0: Length of jeans is random versus H_1: Length of jeans is not random

b.   Open the data file: Use the menu **File>Other Files>Import ASCII Data**; type **c1** for *column* and click *OK*. Give the file name *jeans.dat* and its location. Type the heading **Jeans** for the column of data.

c.   To use the *Runs* test, select **Stat>Nonparametric>Runs Test;** for variables type or double-click *Jeans*; click on *Above and Below* and click on its box and type **33.999** and click *OK*.

1.   [Note that the test value is 34 inches, but for this example we've used 33.999, instead. This is because Minitab considers an exact match like 34 as belonging to the *Above* category, while the text places it in the *Below* category. To force Minitab to match the approach of the book for this problem, use a slightly smaller value than the test value.] The Minitab command you would type if you didn't use the menus is:
MTB > Runs 33.9999   'Jeans'.

d.   The display is shown below. Since the "p-value" or significance level is 0.0055, we would be comfortable in rejecting H_0. The process appears to not be random. Another approach is to note that the number of runs is $U = 8$, and the table critical value for $n_1 = 11$ and $n_2 = 19$ is 9, so since $U_{sample} = 8 < U_L = 9$, reject H_0.
```
Runs Test
Jeans
K =     33.9999
The observed no. of runs =   8
     The expected no. of runs =  14.9333
     11 Observations above K   19 below
              The test is significant at  0.0055
```

USING STATISTICAL PACKAGES
In this section we'll review a number of analysis tools using the EMPSAT.DAT data.

Open the data file: Use the menu **File>Other Files>Import ASCII Data**; type **c1-c29** for *column* and click *OK*. Give the file name *empsat.dat* and its location. Type the heading **Hours** for column c2 and **Rincome** for column c8.

PROBLEM 12.36
Test at α=.05 whether the average number of hours worked differs from 42.

a.   Descriptive statistics: type
```
MTB > Describe 'hours'
```
b.   The display is shown.
```
Descriptive Statistics
Variable        N      Mean    Median    TrMean    StDev   SEMean
Hours          400    45.432   40.000    44.603    10.046   0.502
Variable       Min      Max       Q1        Q3
Hours         28.000   89.000   40.000    50.000
```
c.   Obtain the Stem-and-Leaf display: From the menu
select **Stat>EDA>Stem-and-Leaf**; double-click *Hours* and click *OK*.
d.   If you typed rather than using the menu,
```
MTB > Stem-and-Leaf 'Hours'.
```
e.   The *stem-and-leaf* display is shown.
```
Character Stem-and-Leaf Display
Stem-and-leaf of Hours     N  = 400
Leaf Unit = 1.0

    2     2 89
   21     3 0000012222222222244
   44     3 5555555556666777788888
 (195)    4 0000000000000000000000000000000000000000000000000000000000000000+
  161     4 5555555555555555555555566666778888888888899
  118     5 000000000000000000000000000000000000000001222244
   69     5 5555555555566679
   52     6 000000000000000000000000234
   23     6 5555578
   16     7 0002223
    9     7 556
    6     8 0004
    2     8 99
```

f.   BoxPlot: From the menu select **Graph>Boxplot**; For *Graph Variables* click in the box under *Y* and double-click *Hours*; Select **Annotation>Title** and type the title **BoxPlot for Hours (Ex 12.36)**, click in the box under *Text Size* and type **1** to reduce the title size. Click *OK*. Click on the *Options* button and click on *Transpose X and Y* so the BoxPlot will have the same form as the character-based boxplot. Click *OK* twice.

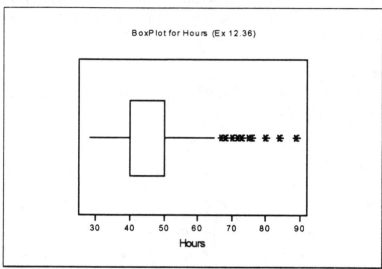

g.   If you typed rather than used the menus,

```
MTB > Boxplot 'Hours';
SUBC>    Transpose;
SUBC>    Box;
SUBC>    Symbol;
SUBC>      Outlier;
SUBC>    Title "BoxPlot for Hours (Ex 12.36)";
SUBC>      TSize 1;
SUBC>    Tick 1;
SUBC>    Tick 2.
MTB > GStd.
MTB > BoxPlot 'Hours'.
```

Character BoxPlot

a.   From the menu select **Graph>Character Graphs>Boxplot** and for *Variable* type **hours** or double-click on *hours* and click on *OK*.

b.   If you typed rather than using the menus,

```
MTB > Gstd.    #to change to character graphs
MTB > BoxPlot 'Hours'.  #generate the Boxplot
MTB > Gpro.      #convert back to Professional graphics.
```

c.   The display is shown.

```
Character Boxplot

                     ----------
         ----------+      I------------ *** ****    *   o    o
                     ----------
         +---------+---------+---------+---------+---------+------Hours
         24        36        48        60        72        84
```

Normal Probability Plot

a. For the data in the *Hours* column, we need to generate normal scores and place them in column c30, and provide a heading. At the Session Window type

```
MTB > nscores 'Hours' c30          #gets normal scores
MTB > name c30 'NormScor'          #types column heading
MTB > gstd.                        #prepares for a character graph
MTB > plot 'Hours' 'NormScor'      #obtains the plot
```

b. The Normal Probability Plot is shown below. This is not a straight line, so there is some amount of deviation from a normal distribution.

Character Plot

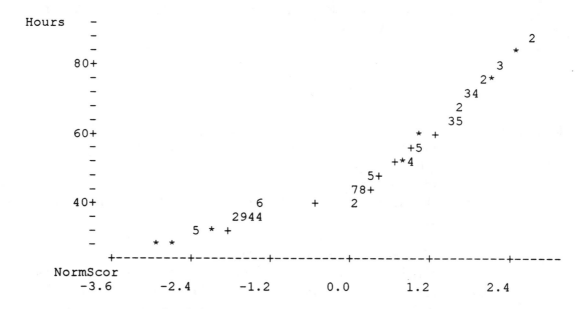

c. To restore to *Professional Graphics*, type
```
MTB > gpro.
```

d. Test the hypotheses: H_0: The average hours worked is 42 versus H_1: The average is not 42; or, H_0: $\mu_x = 42$ versus H_1: $\mu_x \neq 42$.

e. Use the *1-Sample t*: Select the menu **Stat>Basic Statistics>1-Sample t**: for *variables* type or double-click *Hours*; click on *Test Mean* and click on its box and type **42**; For the Alternate (hypothesis) choose *Not Equal* and click *OK*.

f. The Minitab commands you would type if you didn't use the menus are:
```
MTB > TTest 42 'Hours';
SUBC>   Alternative 0.
```

g. The display is shown below. Without additional work it is easy to see that we should reject H_0, since the p-value is 0. This is much smaller than any reasonable α such as 0.05. Also note that the sample t value is 6.83, which is very large (it is far off the axis).

```
T-Test of the Mean
Test of mu = 42.000 vs mu not = 42.000
Variable      N        Mean     StDev   SE Mean        T    P-Value
Hours       400      45.432    10.046     0.502     6.83     0.0000
```

h. We could find the critical t value (although it isn't necessary; we already know the conclusion). For a two-tailed test, the upper critical value is INVCDF$(1-\alpha/2)$ = INVCDF(.975). Type

```
MTB > invcdf .975 ;
SUBC> t 399 df.
```

i. The display is shown below. Since the sample t = 6.83 is much larger than the critical t of 1.9659, we know to reject H_0.

```
Inverse Cumulative Distribution Function
Student's t distribution with 399 d.f.
P( X <= x)              x
   0.9750          1.9659
```

RUNS TEST

a. Is the sample random? The hypotheses are H_0: the sample is random, versus H_1: the sample is not random.

b. To use the *Runs* test, select **Stat>Nonparametric>Runs Test;** for *variables* type or double-click *Hours*; click on *Above and Below* and click on its box and type **42** and click *OK*.

c. If you didn't use the menus, the Minitab command is shown below.

```
MTB > Runs 42.0000   'Hours'.
```

d. The display is shown below. Since the "p-value" or significance level is .1193, we cannot reject H_0, so the sample is likely to be random. Because n is very large, the table for the runs statistic would not be available. A Z-value could be used to compute a sample Z to compare to the critical Z.

```
Runs Test
Hours
K =     42.0000
The observed no. of runs = 181
     The expected no. of runs = 196.1950
   169 Observations above K   231 below
              The test is significant at  0.1193
              Cannot reject at alpha = 0.05
```

TWO-SAMPLE TESTS WITH NUMERICAL DATA

The goal of this chapter is to extend the principles of hypothesis testing to two-sample tests using numerical variables.

POOLED-VARIANCE T TEST FOR DIFFERENCES IN TWO MEANS

The test statistic of interest for this case, where the variances in the two samples are assumed to be equal, is

$$t = \frac{(\overline{X}_1 - \overline{X}_2) - (\mu_1 - \mu_2)}{\sqrt{S_p^2\left(\frac{1}{n_1} + \frac{1}{n_2}\right)}}$$

PROBLEM 13.9

Test at $\alpha = .05$ whether the yield at two banks is the same.

a. If the data aren't current, open the file by using the menu **File>Other Files>Import ASCII Data**; type **c1-c2** for *column* and click *OK*. Give the file name *nybanks.dat* and its location. Type the heading **Yield** for column c1 and **Code** for column c2.

b. The hypotheses are H_0: $\mu_{x1} = \mu_{x2}$ versus H_1: $\mu_{x1} \neq \mu_{x2}$

c. Select the menu **Stat>Basic Statistics>2-Sample t**: Click on *Samples in one column* and double-click *Yield* for the *samples* column and double-click *Code* for the *subscripts* column. The *Alternative* hypothesis is *not equal to*, the confidence is .95, and click on the box *Assume equal variances* (which means *pooling*).

d.   The Minitab commands you would type if you didn't use the menus are:
```
MTB > TwoT 95.0 'Yield' 'Code';
SUBC>    Alternative 0;
SUBC>    Pooled.
```

e.   The display is shown below. Without additional work it is easy to see that we should reject H_0, since the p-value is .0025. This is much smaller than any reasonable α such as 0.05. Also note that the sample t value is -3.52, which is very small (it is off the axis).
```
Two Sample T-Test and Confidence Interval
Twosample T for Yield
Code   N      Mean     StDev    SE Mean
0     10     2.141    0.191     0.060
1     10     2.561    0.326      0.10

95% C.I. for mu 0 - mu 1: ( -0.671,  -0.17)
T-Test mu 0 = mu 1 (vs not =): T= -3.52  P=0.0025  DF=  18
Both use Pooled StDev = 0.267
```

f.   We could find the critical t value (although it isn't necessary; we already know the conclusion). For a two-tailed test, the lower critical value is $INVCDF(\alpha/2) = INVCDF(.025)$. Here there are $n_1 + n_2 - 2 = 10 + 10 - 2 = 18$ degrees of freedom. Type
```
MTB > invcdf .025 ;
SUBC> t 18 df.
```

g.   The display is shown below. Since the sample t = -3.52 is much smaller than the critical t of -2.1009, we know to reject H_0.
```
Inverse Cumulative Distribution Function
Student's t distribution with 18 d.f.
P( X <= x)           x
    0.0250        -2.1009
```

PROBLEM 13.19
This problem uses the same data as problem 13.9, NYBANKS.DAT. The difference is that we do not assume equal variance, so there is no pooling of the variances. The test is referred to as the *separate-variance* t-test. (Note that what is presented here is the Minitab version of the problem. The authors provide a slightly different approach to the same problem.)

a.   Follow the same procedures as 13.9, except for this problem **leave unchecked** the *assume equal variances* box.

b.   If you typed the commands rather than using the menus, you would type the following (note the absence of the keyword *Pooled*).
```
MTB > TwoT 95.0 'Yield' 'Code';
SUBC>    Alternative 0.
```

c.   The display is shown below. Notice that instead of 18 degrees of freedom, as in problem 13.9, there are 14 degrees of freedom. The computation of the degrees of freedom requires a complicated formula (see the text), but Minitab provides this automatically. It is clear that we would reject H_0, since the p-value is .0034. This is much smaller than any reasonable α such as 0.05. Also note that the sample t value is a very small -3.52, as it was previously.

```
Two Sample T-Test and Confidence Interval
Twosample T for Yield
Code    N       Mean    StDev   SE Mean
0       10      2.141   0.191   0.060
1       10      2.561   0.326   0.10

95% C.I. for mu 0 - mu 1: ( -0.676,  -0.16)
T-Test mu 0 = mu 1 (vs not =): T= -3.52  P=0.0034  DF=  14
```

d. We could find the critical t value (although it isn't necessary; we already know the conclusion). For a two-tailed test, the lower critical value is $\text{INVCDF}(\alpha/2) = \text{INVCDF}(.025)$. Here there are 14 degrees of freedom. Type

```
MTB > invcdf .025 ;
SUBC> t 14 df.
```

e. The display is shown below. Since the sample t = -3.52 is much smaller than the critical t of -2.1448, we know to reject H_0.

```
Inverse Cumulative Distribution Function
Student's t distribution with 14 d.f.
P( X <= x)              x
    0.0250          -2.1448
```

f. A careful analyst will perform a number of procedures to try to determine the best decision. To get summary statistics, type the following. Note the *by* keyword, which causes Minitab to treat the first 10 rows of the *Yield* column differently from the second 10 rows, because of the code which was placed in the column *Code*. (The code of 0 pertains to Commercial Banks, and the code of 1 pertains to Savings Banks.) Type

```
MTB > describe 'Yield' ;
SUBC> by 'code'.
```

g. The result is shown below. The *Code* of 0 pertains to the first 10 rows.

```
Descriptive Statistics
Variable   Code     N     Mean   Median  TrMean   StDev   SEMean
Yield       0      10    2.1410  2.1750  2.1512  0.1910  0.0604
            1      10    2.561   2.500   2.549   0.326   0.103

Variable   Code    Min     Max      Q1      Q3
Yield       0     1.8200  2.3800  1.9950  2.3050
            1     2.020   3.200   2.418   2.662
```

h. Similarly, stem-and-leaf results might be obtained.
Select **Stat>EDA>Stem-and-Leaf**; double-click *Yield* and click on *by variable* and double-click on *Code*. If rather than using the menus you typed it:

```
MTB > Stem-and-Leaf 'Yield';
SUBC>  By 'Code'.
```

i.   The display is shown below. The data do not look particularly normal, although this may be because these are small sample sizes.

```
Character Stem-and-Leaf Display
Stem-and-leaf of Yield        Code = 0          N  = 10
Leaf Unit = 0.10
      2     1 89
      5     2 001
      5     2 22333

Stem-and-leaf of Yield        Code = 1          N  = 10
Leaf Unit = 0.10
      1     2 0
      2     2 3
     (6)    2 455555
      2     2
      2     2
      2     3 0
      1     3 2
```

normal probability plot

j.   The data for this problem (*nybanks.dat*) are stacked in one column *Yield*, so that the first 10 rows in *Yield* pertain to *Commercial Banks*, and the next 10 rows in *Yield* pertain to *Savings Banks*. To obtain the two normal probability plots, we need to unstack the data; put it into two separate columns. Select **Manip>Unstack** and for the *Unstack* column double-click on *Yield*; for the *Using subscripts in* double-click on *Code*; for the first block type **c4** and for the second block (the next box) type **c5**.

k.   If you typed rather than used the menu,
```
MTB > Unstack ('Yield') (c4) (c5);
SUBC>    Subscripts 'Code'.
```

l.   Provide headings for these columns: for the first one (c4 in this example) type the heading **Commerc** and for the next one (c5) type **Savings**.

m.   To obtain the Normal Probability Plot: For the data in the *Commerc* column, we need to generate normal scores and place them in column c7, and provide a heading. At the Session Window type
```
MTB > nscores 'Commerc' c7        #gets normal scores
MTB > name c7 'NormComm'          #types column heading
MTB > gstd.                       #prepares for character graph
MTB > plot 'Commerc' 'NormComm'   #obtains the plot
```

n.   The Normal Probability Plot is shown below. This is not a straight line, so there is some amount of deviation from a normal distribution.

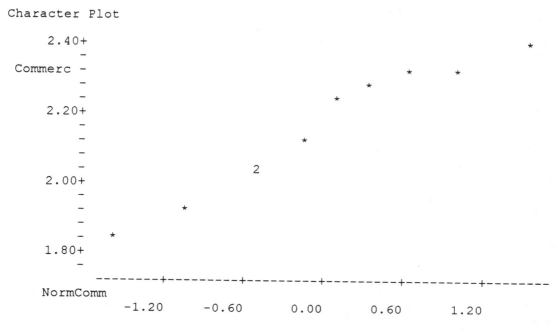

```
Character Plot

       2.40+                                                        *
            -
 Commerc   -                                          *        *
            -
            -                                      *
       2.20+                                   *
            -
            -                         *
            -                  2
       2.00+
            -
            -          *
            -
            -    *
       1.80+
            -
            --------+---------+---------+---------+---------+--------
 NormComm
            -1.20      -0.60      0.00       0.60       1.20
```

Normal Probability Plot

o.   For the data in the *Savings* column, we need to generate normal scores and place them in column c8, and provide a heading. At the Session Window type

```
MTB > nscores 'Savings' c8        #gets normal scores
MTB > name c8 'NormSave'          #types column heading
MTB > gstd.                       #prepares for character graph
MTB > plot 'Savings' 'NormSave'   #obtains the plot
```

p.   The Normal Probability Plot is shown below. This is not a straight line, so there is some amount of deviation from a normal distribution.

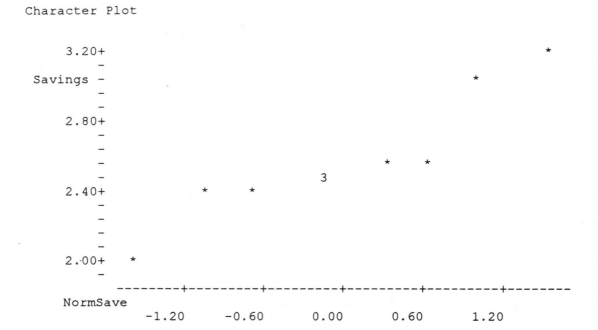

Character Plot

WILCOXON RANK SUM TEST FOR DIFFERENCES IN TWO MEDIANS

This nonparametric test is quite powerful for testing for differences between the medians of two populations.

The procedure is to rank all $n = n_1 + n_2$ values as one group. Then compute the statistic

$T_1 = \Sigma$(ranks of the smaller sample),

and if the sample sizes are the same, choose either sample.

PROBLEM 13.30

This problem uses the same data as problem 13.9 and problem 13.19, NYBANKS.DAT.

Test at $\alpha=.05$ whether there is a difference in median yield at two banks.

a.   If the data aren't currently loaded, open the file by using the menu **File>Other Files>Import ASCII Data**; type **c1-c2** for *column* and click *OK*. Give the file name *nybanks.dat* and its location. Type the heading **Yield** for column c1 and **Code** for column c2.

b.   First rank the entire group of 20 values and store the ranks in c3. Type the heading for c3, **Ranks**.

c.   Select **Manip>Rank** and for *Rank data in* double-click on *Yield*; for *Store ranks in* double-click on *Ranks*.

d.   If you typed rather than used the menu the command would be
```
MTB > Rank 'Yield' 'Ranks'.
```

e.   Since the samples are the same size, we'll base our statistic on the first sample. To compute the sum of the ranks of just the first sample, copy just those ranks to another column. Type the heading **RnkComm** (meaning the ranks for the commercial banks) for column c4.

f.   The goal is to use the subscripts in the column *Code* (which are 0 for the first sample and 1 for the second sample) to pick out only the ranks for the first sample. At the Session Window type
```
MTB > copy 'Ranks' 'RnkComm';
SUBC> use 'code' = 0.
MTB > sum 'RnkComm'
```

g.   The display is as follows, which is the sum of the ranks for the first sample; sample T = 62.5. This may be compared to the critical T value from the table in the text for $n_1 = 10$ and $n_2 = 10$, which is critical T = 78. Since (sample T = 62.5 is < critical T = 78), reject H_0. Therefore, the medians in the two banks seem to be different, since we rejected H_0 of no difference in the median yields.
```
Column Sum
Sum of RnkComm =          62.500
```

h.   Similarly, a sample Z value may be computed by

$$Z = \frac{T_1 - \frac{n_1(n+1)}{2}}{\sqrt{\frac{n_1 n_2 (n+1)}{12}}} \qquad Z = \frac{62.5 - \frac{10(20+1)}{2}}{\sqrt{\frac{10(10)(20+1)}{12}}} = -3.213 \ .$$

i.   The lower critical value for a two-tailed test and $\alpha=.05$ may be computed by $F(\alpha/2 = 0.025)$. Type
```
MTB > invcdf .025;
SUBC> normal mean 0 stdev 1 .
```

j.   The display is shown below. Since the sample z = -3.213 < critical z = -1.96, we are led to reject H_0.
```
Inverse Cumulative Distribution Function
Normal with mean = 0 and standard deviation = 1.00000
   P( X <= x)              x
      0.0250          -1.9600
```

F Test for Differences in Two Variances
The goal is to test for a difference in variability, which plays a role in determining whether, for instance, to use the *pooled-variance* t-test or *separate-variance* t-test. The ratio of sample variances follows an F distribution: $F = \dfrac{S_1^2}{S_2^2}$

PROBLEM 13.42
This problem uses the same data as problems 13.9, 13.19, and 13.30: NYBANKS.DAT.
Test at $\alpha=.05$ whether there is a difference in the variances of yield at two banks.

a.   If the data aren't currently loaded, open the file by using the menu **File>Other Files>Import ASCII Data**; type c1-c2 for *column* and click *OK*. Give the file name *nybanks.dat* and its location. Type the heading **Yield** for column c1 and **Code** for column c2.

b. The hypotheses are $H_0: \sigma_1^2 = \sigma_2^2$ versus $H_1: \sigma_1^2 \neq \sigma_2^2$.

c. Obtain the sample standard deviations for the two samples. Recall that the values for both samples are in one column *Yield*, and the column *Code* indicates which rows pertain to which bank. Type
```
MTB > describe 'yield';
SUBC> by 'code'.
```

d. The display is as shown.
```
Descriptive Statistics
Variable    Code        N      Mean    Median    TrMean     StDev    SEMean
Yield          0       10    2.1410    2.1750    2.1512    0.1910    0.0604
               1       10     2.561     2.500     2.549     0.326     0.103

Variable    Code      Min       Max        Q1        Q3
Yield          0    1.8200    2.3800    1.9950    2.3050
               1     2.020     3.200     2.418     2.662
```

e. There is no direct Minitab command for computing the F for this situation, so we'll do the computations using the *Let* command. Compute the F Statistic: Type (the #comments are optional)
```
MTB > let k1 = .1910          #stdev commerce
MTB > let k2 = .326           #stdev savings
MTB > let k3 = k1**2/k2**2
MTB > print k3
```

f. The display is as shown. The F statistic is 0.343267.
```
Data Display
K3        0.343267
```

g. Next compute the p-value by finding the CDF for the sample F value. Recall that the numerator df = 10-1 = 9, as are the denominator df. K3 is the F statistic just computed. Type
```
MTB > cdf k3;          #compute p-value
SUBC> f 9 and 9 df.
```

h. The display is shown. The left tail area for the sample F of .343 is .0635, and the p-value is twice this value, since this is a two-tailed test: p-value = 2*.0635 = 0.1270. On this basis, we could not reject H_0, so the conclusion is that the population variances are about the same.
```
Cumulative Distribution Function
F distribution with 9 d.f. in numerator and 9 d.f. in denominator
     x          P( X <= x)
   0.3433          0.0635
```

i. The tabled or critical F values could be computed by:
```
MTB > invcdf .975 ;
SUBC> f 9 and 9 df.
MTB > invcdf .025 ;
SUBC> f 9 and 9 df.
```

j.   The result is shown below. Using the critical values we are again led to not reject H_0, since lower critical F = 0.2484 < sample F = 0.343 < upper critical F = 4.026.

```
Inverse Cumulative Distribution Function
F distribution with 9 d.f. in numerator and 9 d.f. in denominator
P( X <= x)              x
    0.9750          4.0260
P( X <= x)              x
    0.0250          0.2484
```

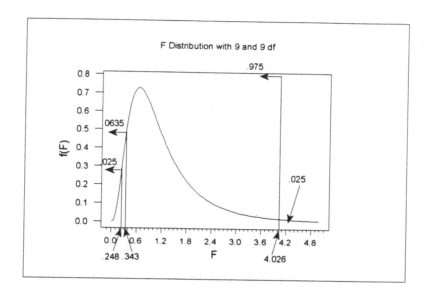

RELATED SAMPLES

t Test for the Mean Difference

When the data are paired or matched, then the distribution of interest is the distribution of differences. Minitab requires that the user compute the differences between the two samples, and then Minitab will perform a one-tailed test on the differences.

PROBLEM 13.61

Test at $\alpha = .05$ whether there is a difference between the appraisal values of two agents on the same house.

a.   If the data aren't currently loaded, open the file by using the menu **File>Other Files>Import ASCII Data**; type **c1-c3** for *column* and click *OK*. Give the file name *lire.dat* and its location. Type the heading **House** for column c1, **Agent1** for column c2, **Agent2** for column c3, and **Diff** for column c4.

b.   The hypotheses are H_0: $\mu_D = 0$ versus H_1: $\mu_D \neq 0$, where $\mu_D = \mu_1 - \mu_2$.

c.   Compute the difference in the appraisal values for the two agents and store the differences in the column *Diff*. Type
       ```
       MTB > let 'diff' = 'agent1' - 'agent2'
       ```

d. Select **Stat>Basic Statistics>1-Sample t**, click on *Test mean* and make sure the mean value in the box is 0.0. Make sure the *Alternative* is *Not equal*. Leave the confidence level at .95. If you typed rather than used the menu:
```
MTB > TTest 0.0 'Diff';
SUBC>    Alternative 0.
```

e. The display is shown. The t value of -0.30 is very small. The p-value of 0.77 for the two-tailed test is very high, so we are led to not reject H_0. There doesn't seem to be a difference in the appraisals from the two agents.
```
T-Test of the Mean
Test of mu = 0.000 vs mu not = 0.000
Variable      N      Mean    StDev   SE Mean       T    P-Value
Diff         12    -0.167    1.904     0.550    -0.30      0.77
```

f. The tabled or critical t values could be computed by:
```
MTB > invcdf .975 ;
SUBC> t 11 df.
MTB > invcdf .025 ;
SUBC> t 11 df.
```

g. The display is shown. Therefore
    lower $t_{critical}$ -2.201 < sample t = -0.30 < upper $t_{critical}$ 2.201, which again leads to the decision to not reject H_0.
```
Inverse Cumulative Distribution Function
Student's t distribution with 11 d.f.
P( X <= x)           x
    0.9750        2.2010
P( X <= x)           x
    0.0250       -2.2010
```

Wilcoxon Signed-Ranks Test for the Median Difference
The nonparametric or distribution free approach for testing matched samples is the Wilcoxon signed-ranks test for the median difference.

PROBLEM 13.66
Test at $\alpha=.05$ whether the amount of taxes paid if firm-prepared is less than if self-prepared.

a. If the data aren't currently loaded, open the file by using the menu **File>Other Files>Import ASCII Data**; type **c1-c3** for *column* and click *OK*. Give the file name *TaxRet.dat* and its location. Type the heading **WhichOne** for column c1, **Firm** for column c2, **Self** for column c3, and **Diff** for column c4.

b. The hypotheses are H_0: $\mu_D \geq 0$ versus H_1: $\mu_D < 0$, where $\mu_D = \mu_1 - \mu_2$. Since the claim is that the *firm-prepared* taxes are lower, the differences of interest will be negative.

c. Compute the differences in the taxes for the two approaches and store in the column *Diff*. Type
```
MTB > let 'diff' = 'firm' - 'self'
```

d.   Now use the Wilcoxon test for the median. Select the menu **Stat>Nonparametrics>1-Sample Wilcoxon**: for *variables* type or double-click *Diff*; click on *Test median* and click on its box and type **0**; For the Alternate (hypothesis) choose *Less than* and click *OK*. The Minitab commands you would type if you didn't use the menus are shown below.

```
MTB > WTest 0.0  'Diff';
SUBC>    Alternative -1.
```

e.   The Minitab display is shown. The Wilcoxon statistic is 35, the sample median is 20.0, and the p-value is 0.894. With a p-value this large we would be led to not reject H_0. There is no evidence that the median taxes for returns prepared by firms are less than those prepared by the individuals.

```
Wilcoxon Signed Rank Test
TEST OF MEDIAN = 0.000000 VERSUS MEDIAN N.E. 0.000000
N FOR    WILCOXON                ESTIMATED
          N    TEST  STATISTIC   P-VALUE     MEDIAN
Diff     11    11       35.0      0.894       20.00
```

ANOVA AND OTHER C-SAMPLE TESTS WITH NUMERICAL DATA

This chapter introduces the concepts of experimental design, using the completely randomized design model and the one-way ANOVA procedure.

Given the assumptions of normality, randomness, and that the variances of the groups are equal, ANOVA tests *means* by analyzing *variances*: by looking at the variance estimates from several different estimators.

PROBLEM 14.7
This problem pertains to average math scores using three different curricula.

a. If the data aren't currently loaded, open the file by using the menu **File>Other Files>Import ASCII Data**; type **c1-c2** for *column* and click *OK*. Give the file name *Curr.dat* and its location. Type the heading **Subscr** for column c1 and **Scores** for column c2.

Hartley's F test for Homogeneity of Variance
Before performing some tests, it would be useful to test whether the variances can be assumed to be equal.

a. The sample F is computed by $F_{max} = \dfrac{S^2_{max}}{S^2_{min}}$, where S^2_{max} is the largest variance and S^2_{min} is the smallest variance.

b. The hypotheses are H_0: $\sigma^2_1 = \sigma^2_2 = \sigma^2_3$ versus H_1: At least one $\sigma^2_i \neq \sigma^2_j$. Is the assumption of equal variance reasonable? Test at $\alpha = .05$.

119

c. Compute the variances for the three samples. Note that all results are in one column, so the subscripts column must be used to pick out separate samples. Type
```
MTB > describe 'scores';
SUBC> by 'subscr' .
```
d. The display is shown below. The smallest standard deviation is 7.13, and the largest is 10.24 .

```
Descriptive Statistics
Variable   Subscr      N     Mean   Median   TrMean    StDev   SEMean
Scores        1        7    82.14    81.00    82.14     7.99     3.02
              2        9    68.44    70.00    68.44     7.13     2.38
              3        8    71.75    70.00    71.75    10.24     3.62

Variable   Subscr    Min      Max       Q1       Q3
Scores        1    74.00    97.00    74.00    87.00
              2    58.00    80.00    62.50    74.00
              3    62.00    92.00    63.25    78.75
```

e. Compute F_{max}: type
```
MTB > let k1 = (10.24**2)/(7.13**2)          #find Fmax
MTB > print k1
```
f. The display is shown. $F_{max} = 2.06263$. Compare this to the critical tabled value (Table E8 in the text, not the usual F table). The degrees of freedom are: number of classes = 3; integer part of $\dfrac{7+9+8}{3}$ -1 = 7, so $F_{max(3,7)} = 6.94$ for the upper 5%. Then, since the sample $F_{max} = 2.06263 <$ critical $F_{max} = 6.94$, we are led by the sample to not reject H_0, so the assumption of equal variance seems reasonable.
```
Data Display
K1         2.06263
```

One-Way ANOVA F Test for Differences in Means

a. The hypotheses are H_0: $\mu_1 = \mu_2 = \mu_3$ versus H_1: At least one $\mu_i \neq \mu_j$. Do the three curricula provide different test results? Test at α=.05.

b. Select **Stat>ANOVA>One Way** and for the *Response* box double-click on *Scores* and for the *Factors* box double-click on *Subscr*. Click on the *Comparisons* button, click on *Tukey's family error rate* and make sure the box shows *5*. Click on *OK*.

c. If you typed rather than used the menu,
```
MTB > Oneway 'Scores' 'Subscr';
SUBC>    Tukey 5.
```
d. The display is shown. The sample F value is 5.36. Since the p-value of 0.013 is so small, we are led by the sample to reject H_0: at least one curriculum is different from the others.

```
One-Way Analysis of Variance
Analysis of Variance on Scores
Source      DF        SS        MS        F         p
Subscr       2     777.4     388.7     5.36     0.013
Error       21    1522.6      72.5
Total       23    2300.0
```

```
                                      Individual 95% CIs For Mean
                                      Based on Pooled StDev
   Level     N      Mean     StDev   --+---------+---------+---------+----
     1       7     82.143     7.988                      (--------*------)
     2       9     68.444     7.126   (-------*------)
     3       8     71.750    10.236      (-------*-------)
                                      --+---------+---------+---------+----
 Pooled StDev =     8.515            64.0      72.0      80.0      88.0
```

```
   Tukey's pairwise comparisons
   Family error rate = 0.0500
   Individual error rate = 0.0200
   Critical value = 3.56
   Intervals for (column level mean) - (row level mean)
                        1            2
          2          2.90
                    24.50

          3         -0.70       -13.72
                    21.49         7.11
```

e. To compare the sample F = 5.36 to the critical, tabled value, compute the degrees of freedom as c-1 = 3-1 = 2 and n-c = 7+9+8-3 = 21, and type
    MTB > **invcdf .95;**
    SUBC> **f 2 21 df .**

f. The display is shown. For $\alpha=.05$, the sample F = 5.36 > the critical $F_{(2,21)}$ = 3.4668, so the decision would be to reject H_0.
    ```
    Inverse Cumulative Distribution Function
    F distribution with 2 d.f. in numerator and 21 d.f. in denominator
    P( X <= x)              x
         0.9500          3.4668
    ```

g. For the *Tukey-Kramer* procedure, refer to output listed above. This procedure is useful in identifying which of the three procedures is different. Since there are three categories or groups, there are 3*2/2 = 3 possible comparisons to be made. The *Tukey-Kramer* critical value is Q=3.56 (which is the tabled value from Table E12). From the output, curricula 1 and 2 are different, but for 2 and 3 and for 1 and 3 there is no difference: if the interval from the output contains 0, there is no difference in the two factors being compared. Curriculum 1 does seem to do the best; it has a higher mean.

Kruskal-Wallis Rank Test for Differences in Medians
This is the distribution-free alternative to the *One-way Analysis*. Note that if two or more observations are tied, the average rank is assigned to each observation. The decision rule is:
    Reject H_0 if the sample H score > the critical χ^2 with c-1 df.

PROBLEM 14.16
(Using the data from Problem 14.6): This problem pertains to average math scores using three different curricula.

a.   If the data aren't currently loaded, open the file by using the menu **File>Other Files>Import ASCII Data**; type **c1-c2** for *column* and click *OK*. Give the file name *Curr.dat* and its location. Type the heading **Subscr** for column c1 and **Scores** for column c2.

b.   The hypotheses are H_0: $M_1 = M_2 = M_3$ versus H_1: At least one $M_i \neq M_j$. Do the three curricula provide different test results? From Problem 14.7 we concluded that there was a difference in means. Now test at $\alpha = .05$ whether there is a difference in medians.

c.   Select **Stat>Nonparametric>Kruskal-Wallis** and for the *Response* box double-click on *Scores* and for the *Factors* box double-click on *Subscr*. Click on *OK*.

d.   If you typed rather than used the menu,
```
MTB > Kruskal-Wallis 'Scores' 'Subscr'.
```

e.   The display is shown. The sample H value is 8.61. Since the p-value of 0.014 is so small, we are led by the sample to reject H_0: at least one curriculum is different from the others.
```
Kruskal-Wallis Test
LEVEL     NOBS     MEDIAN    AVE. RANK    Z VALUE
  1          7      81.00        19.0       2.89
  2          9      70.00         9.0      -1.88
  3          8      70.00        10.8      -0.86
OVERALL     24                   12.5
H = 8.61   d.f. = 2   p = 0.014
H = 8.65   d.f. = 2   p = 0.013 (adjusted for ties)
```

f.   To compare the sample H = 8.61 to the critical, tabled χ^2 value, note that the degrees of freedom = c-1 = 3-1 = 2. Type
```
MTB > invcdf .95;
SUBC> chisq 2 df .
```

g.   The display is shown. For $\alpha = .05$, the sample H = 8.61 > the critical χ^2 = 5.9915, so we would be led by the sample to reject H_0.
```
Inverse Cumulative Distribution Function
Chisquare with 2 d.f.
P( X <= x)          x
    0.9500       5.9915
```

Randomized Block Model

Previously, we used the t test or Wilcoxon signed-ranks for comparing two related groups. Now we wish to look at more than two related groups. Note that for the *TWOWAY* command, the design must be balanced; that is, the same number of observations in each group.

PROBLEM 14.34

This problem pertains to weight loss using three different products.

a.   If the data aren't currently loaded, open the file by using the menu **File>Other Files>Import ASCII Data**; type **c1-c3** for *column* and click *OK*. Give the file name

Diet.dat and its location. Type the heading **RowFact** for column c1, **ColFact** for column c2, and **Response** for column c3.

b. The hypotheses are H_0: $\mu_1 = \mu_2 = \mu_3$ versus H_1: At least one $\mu_i \neq \mu_j$. Do the three diet products provide different weight loss results? Test at $\alpha = .05$.

c. Select **Stat>ANOVA>Two Way** and for the *Response* box double-click on *Response*; for the *Row Factor* box double-click on *RowFact* ; for the *Column Factor* box double-click on *ColFact*. Click on the *Display means* boxes for both.

d. If you typed rather than used the menu,
```
MTB > Twoway 'Response' 'RowFact' 'ColFact';
SUBC>   Means 'RowFact' 'ColFact'.
```

e. The display is shown.

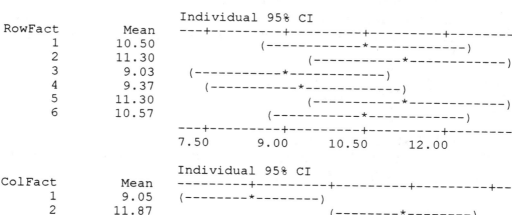

```
Two-way Analysis of Variance
Analysis of Variance for Response
Source         DF         SS        MS
RowFact         5      13.72      2.74
ColFact         2      24.27     12.13
Error          10      21.77      2.18
Total          17      59.76
```

f. Note that the sample $F = \dfrac{MS_{ColFact}}{MS_{error}} = \dfrac{12.13}{2.18} = 5.564$. Minitab may be used to compute the sample F. Type
```
MTB > let k1=12.13/2.18
MTB > print k1
```

g. The display is shown.
```
Data Display
K1          5.56422
```

h. The sample F = 5.564 should be compared with a critical, Table F value with c-1 = 3-1 = 2 and (r-1)(c-1) = 10 df. Type
```
MTB > invcdf .95;
SUBC> f 2 10 df.
```

i.   The display is shown. Since sample $F = 5.564 >$ critical, tabled $F = 4.1029$, reject H_0. There does seem to be a difference in the effectiveness of the weight-loss products.

```
Inverse Cumulative Distribution Function
F distribution with 2 d.f. in numerator and 10 d.f. in denominator
  P( X <= x)              x
     0.9500            4.1029
```

Comparing Randomized Block Design to the One-Way (Completely Randomized Design)

The computation is $RE = \dfrac{(r-1)MSBL + r(c-1)MSE}{(rc-1)MSE} = \dfrac{(6-1)*2.74 + 6*(3-1)*2.18}{(6*3-1)*2.18}$.

a.   (Note that it would likely be easier to use a calculator in this case; this Minitab computation is provided only for your information.)

b.   At the Session Window type

```
MTB > let k1 = ( (6-1)*2.74 + 6*(3-1)*2.18 )/((6*3-
      1)*2.18)
```

c.   The result is displayed. Therefore the efficiency of blocking is $RE = 1.07555$, meaning that it would take 1.08 times as many observations to equal blocking efficiency. In other words, blocking is efficient in the sense of requiring a smaller sample size to achieve the same general conclusion as you would without the blocking.

```
MTB > print k1
Data Display
K1          1.07555
```

Friedman Rank Test for Differences in c Medians

This is the distribution-free analog to the randomized block design. Reject H_0 if the sample H score > the critical χ^2 with c-1 df.

PROBLEM 14.43

(Using the data from Problem 14.35): This problem pertains to faculty ratings.

a.   If the data aren't currently loaded, open the file by using the menu **File>Other Files>Import ASCII Data**; type **c1-c3** for *column* and click *OK*. Give the file name *Rating.dat* and its location. Type the heading **RowFact** for column c1, **ColFact** for column c2, and **Response** for column c3.

b.   The hypotheses are H_0: $M_1 = M_2 = M_3$ versus H_1: At least one $M_i \neq M_j$. Do the three course levels provide different evaluation results?

c.   Select **Stat>Nonparametric>Friedman** and for the *Response* box double-click on *Response*; for the *Treatment* box double-click on *ColFact*; for the *Blocks* box double-click on *RowFact*. Click on *OK*.

d.   If you typed rather than used the menu,
```
MTB > Friedman 'Response' 'ColFact' 'RowFact'.
```
e.   The display is shown. The sample S value is 8.60. Since the p-value of 0.014 is so small, we are led by the sample to reject H_0: at least one course type is different from the others.
```
Friedman Test
Friedman test of Response by ColFact blocked by RowFact
S = 8.60   d.f. = 2   p = 0.014
Est.    Sum of
  ColFact        N     Median     RANKS
        1       10     4.1450      26.0
        2       10     4.0050      21.0
        3       10     3.6550      13.0
Grand median  =     3.9350
```

f.   To compare the sample F_R = 8.60 (shown as S = 8.60 in the Minitab output) to the critical, tabled χ^2 value, note that the degrees of freedom = c-1 = 3-1 = 2. Type
```
MTB > invcdf .95;
SUBC> chisq 2 df .
```
g.   The display is shown. For α=.05, the sample F_R = 8.60 > the critical χ_2^2 = 5.9915, so we would be led by the sample to reject H_0.
```
Inverse Cumulative Distribution Function
Chisquare with 2 d.f.
P( X <= x)              x
     0.9500          5.9915
```

The Factorial Design Model and Two-Way Analysis of Variance

PROBLEM 14.48
This problem pertains to the strength of wool in terms of two factors: operators and machines.

a.   If the data aren't currently loaded, open the file by using the menu **File>Other Files>Import ASCII Data**; type **c1-c3** for *column* and click *OK*. Give the file name *Breaking.dat* and its location. Type the heading **RowFact** for column c1, **ColFact** for column c2, and **Response** for column c3.
b.   There is a set of hypotheses for each factor. The general form is H_0: $\mu_{1..} = \mu_{2..} = \mu_{3..}$ versus H_1: At least one $\mu_{i..} \neq \mu_{j..}$. Test each factor at α=.05.
c.   Select **Stat>ANOVA>Two Way** and for the *Response* box double-click on *Response*; for the *Row Factor* box double-click on *RowFact* ; for the *Column Factor* box double-click on *ColFact*. Click on the *Display means* boxes for both.
d.   If you typed rather than used the menu,
```
MTB > Twoway 'Response' 'RowFact' 'ColFact';
SUBC>   Means 'RowFact' 'ColFact'.
```

e.   The display is shown.
```
Two-way Analysis of Variance
Analysis of Variance for Response
Source        DF        SS        MS
RowFact        3      301.11    100.37
ColFact        2      329.39    164.69
Interaction    6       86.39     14.40
Error         24       90.67      3.78
Total         35      807.56
```

```
                            Individual 95% CI
RowFact      Mean    ---------+---------+---------+---------+--
   1        112.00                                   (----*----)
   2        110.56                             (----*-----)
   3        104.33      (----*-----)
   4        109.56                     (----*-----)
                    ---------+---------+---------+---------+--
                       105.00    107.50    110.00    112.50
```

```
                            Individual 95% CI
ColFact      Mean    --+---------+---------+---------+---------
   1        113.08                                  (---*----)
   2        105.75      (----*----)
   3        108.50             (----*----)
                    --+---------+---------+---------+---------
                    105.00    107.50    110.00    112.50
```

Test for *Operator*

f.   Test at $\alpha = .05$ whether there is an effect due to the operator (the rows). For the rows,

the sample $F = \dfrac{MS_{RowFact}}{MS_{error}} = \dfrac{100.37}{3.78} = 26.5529$. The sample F may be computed by the

following commands.
```
MTB > let k1=100.37/3.78
MTB > print k1
```

g.   The display is shown.
```
Data Display
K1          26.5529
```

h.   The sample F may be compared to the critical F from the table. The degrees of freedom are r-1 = 4-1 = 3 and rc(n'-1) = 4*3*2 = 24, where n' = number of replications for each cell. Therefore, type
```
MTB > invcdf .95;
SUBC> f 3 24 df.
```

i.   The display is shown. Since sample F = 26.5529 > critical, tabled F = 3.0088, reject H_0. There does seem to be a difference in the effectiveness of the machine operators.
```
F distribution with 3 d.f. in numerator and 24 d.f. in denominator
  P( X <= x)            x
     0.9500         3.0088
```

Test *machine*

j. Test at $\alpha=.05$ whether there is an effect due to the machine (the columns). For the columns, the sample $F = \dfrac{MS_{ColFact}}{MS_{error}} = \dfrac{164.69}{3.78} = 46.5688$. The sample F may be computed by the following commands.

```
MTB > let k1=164.69/3.78
MTB > print k1
```

k. The display is shown.

```
Data Display
K1        46.5688
```

l. The sample F may be compared to the critical F from the table. The degrees of freedom are c-1 = 3-1 = 2 and rc(n'-1) = 4*3*2 = 24. Therefore, type

```
MTB > invcdf .95;
SUBC> f 2 24 df.
```

m. The display is shown. Since sample F = 46.5688 > critical, tabled F = 3.4028, reject H_0. There does seem to be a difference in the effectiveness of the machine.

```
F distribution with 2 d.f. in numerator and 24 d.f. in denominator
  P( X <= x)            x
    0.9500          3.4028
```

Test *interaction*

n. Test at $\alpha=.05$ whether there is an effect due to the interaction between the operators (the rows) and the machines (the columns). For the interaction, the sample $F = \dfrac{MS_{Interact}}{MS_{error}} = \dfrac{14.40}{3.78} = 3.8095$. The sample F may be computed by the following commands.

```
MTB > let k1=14.40/3.78
MTB > print k1
```

o. The display is shown.

```
Data Display
K1        3.8095
```

p. The sample F may be compared to the critical F from the table. The degrees of freedom are (r-1)(c-1) = 3*2 = 6 and rc(n'-1) = 4*3*2 = 24. Therefore, type

```
MTB > invcdf .95;
SUBC> f 6 24 df.
```

q. The display is shown. Since sample F = 3.8095 > critical, tabled F = 2.5082, reject H_0. There does seem to be an interaction between operators and machines which cannot be separated into a simple operator effect or a simple machine effect.

```
F distribution with 6 d.f. in numerator and 24 d.f. in denominator
  P( X <= x)            x
    0.9500          2.5082
```

r. To obtain a table of the means and the data, Select **Stat>Tables>Cross Tabulation**; for *Classification variables* double-click on *RowFact* and *ColFact*; click on the *Summaries* button, and for *Associated variables* double-click on *Response* and for *Display* click the boxes for *Means* and *Data*.

s.   If you typed rather than used the menus,
```
MTB > Table 'RowFact' 'ColFact';
SUBC>   Means 'Response';
SUBC>   Data 'Response'.
```

t.   The table is shown below. (The first row in each block of 4 is the average for each cell. For instance, (115+115+119)/3=116.33, which is the first entry in the table and it is the average of the 3 values below it.)

```
Tabulated Statistics
ROWS: RowFact        COLUMNS: ColFact
             1          2          3        ALL

    1     116.33     111.00     108.67     112.00
          115.00     111.00     109.00       --
          115.00     108.00     110.00
          119.00     114.00     107.00


    2     115.00     104.33     112.33     110.56
          117.00     105.00     110.00       --
          114.00     102.00     113.00
          114.00     106.00     114.00


    3     108.33     101.33     103.33     104.33
          109.00     100.00     103.00       --
          110.00     103.00     102.00
          106.00     101.00     105.00


    4     112.67     106.33     109.67     109.56
          112.00     105.00     108.00       --
          115.00     107.00     111.00
          111.00     107.00     110.00
  ALL     113.08     105.75     108.50     109.11
            --         --         --         --
CELL CONTENTS --
          Response:MEAN
                   DATA
```

HYPOTHESIS TESTING WITH CATEGORICAL DATA

This chapter extends the basic principles of hypothesis-testing to situations involving categorical variables.

One-Sample Z test for the Proportion

Here the goal is to use the sample proportion $p_s = \dfrac{X}{n} = \dfrac{\text{Number of successes in sample}}{\text{sample size}}$ to test some hypothesis regarding the population proportion p. Then the test statistic is

$$Z = \frac{p_s - p}{\sqrt{\dfrac{p(1-p)}{n}}} \, .$$

PROBLEM 15.5
What proportion of car buyers would have purchased an air bag? The assumed proportion is 0.30. A sample of n = 200 revealed that 79 would have purchased it. Test this at $\alpha = .10$ level of significance.

     a.   The hypotheses are H_0: p = .30 versus H_1: p $\neq$ .30 .

b. Let Minitab determine the correct Z values (which could be obtained from a table.) The two critical values are for F(.05) and F(.95). The upper critical Z is for F(.95). At the Session Window type

```
MTB > invcdf .95;
SUBC> normal mu 0 stdev 1.
```

c. The display is shown below. The upper critical Z value is 1.6449, and because of symmetry the lower critical Z value is -1.6449, which are the test values.

```
Inverse Cumulative Distribution Function
Normal with mean = 0 and standard deviation = 1
P( X <= x)            x
0.9500           1.6449
```

d. Now to decide whether the sample provides sufficient support to reject H_0, we compare the test Z-value with the sample Z-value. If the sample result converted to a Z-score is less than -1.6449 or more than 1.6449, this will be sufficient evidence to reject H_0 in favor H_1. Transform the sample result to a Z-value by typing

```
MTB > let k1 = (79/200-.30)/sqrt(.30*.70/200)
MTB > print k1
```

e. The result is shown below. Since the sample Z value of 2.93 is greater than 1.6449, then one is lead by this sample result to reject H_0. Therefore the conclusion is that the proportion which would have purchased this option is not .30 . (If fact, it is probably more than 0.30 .)

```
Data Display
K1        2.93176
```

Compute the p-value

f. The p-value is the tail area from the sample Z value. Here the sample Z value (the sample p_s value converted to a Z) was 2.93. The tail area for z = 2.93 is F(2.93). Type

```
MTB > cdf 2.93;
SUBC> normal mu 0 stdev 1 .
```

g. The result is shown below. The tail area for the sample Z of 2.93 is 0.0017. Since this was a two-tailed test, the p-value is twice this tail area, so the p-value is 0.0034 . Since this represents a small risk of a Type I error if the null hypothesis is rejected, we were right to reject H_0 in the previous problem. The p-value is much smaller than a typical α level of risk of 0.05. Since the risk of an error when rejecting H_0 is 0.0034 < 0 .05, it is reasonable to reject H_0.

```
Cumulative Distribution Function
Normal with mean = 0 and standard deviation = 1.00000
     x        P( X <= x)
  2.9300        0.9983
```

Z Test for Differences in Two Proportions (Independent Samples)

Here the goal is to use the two sample proportions P_{s_1} and P_{s_2} to test some hypothesis regarding the difference in two population proportions p_1 and p_2. Then the test statistic is approximately

$$Z = \frac{(p_{s_1} - p_{s_2}) - (p_1 - p_2)}{\sqrt{\bar{p}(1-\bar{p})\left(\frac{1}{n_1} + \frac{1}{n_2}\right)}} \ , \text{ where } \bar{p} = \frac{X_1 + X_2}{n_1 + n_2} \ .$$

PROBLEM 15.13

Is there a difference in the proportion of accountants who found Company A's reports *understandable* versus the proportion of accountants who found Company B's reports *understandable*? In a sample of 50 accountants, 17 found Company A's reports understandable. In a separate sample of 50 accountants, 23 found Company B's reports understandable. Test the hypotheses at $\alpha=.10$ level of significance.

   a.   The hypotheses are H_0: $p_1 - p_2 = 0$ versus H_1: $p_1 - p_2 \neq 0$.
   b.   Let Minitab determine the correct Z values (which could be obtained from a table.) The two critical values are for F(.05) and F(.95). The upper critical Z is for F(.95). At the Session Window type

```
MTB > invcdf .95;
SUBC> normal mu 0 stdev 1.
```

   c.   The display is shown below. The upper critical Z value is 1.6449, and because of symmetry the lower critical Z value is -1.6449, which are the test values.

```
Inverse Cumulative Distribution Function
Normal with mean = 0 and standard deviation = 1
P( X <= x)            x
0.9500             1.6449
```

   d.   Now to decide whether the sample provides sufficient support to reject H_0, we compare the test Z-value with the sample Z-value. If the sample result converted to a Z-score is less than -1.6449 or more than 1.6449, this will be sufficient evidence to reject H_0 in favor H_1. Transform the sample result to a Z-value. Since the formula is complicated, we'll do it in several parts. Type

```
MTB > let k1 = (17 + 23)/ (50 + 50)
MTB > let k2 = (17/50 - 23/50)/sqrt(k1*(1-k1)*(1/50 +
      1/50) )
MTB > print k1, k2
```

   e.   The result is shown below. Since the sample Z value of -1.22474 is greater than -1.6449 (it is inside the acceptance region), then one is lead by this sample result to not reject H_0. Therefore the conclusion is that there is no difference in the proportions of analysts which thought the two annual reports were understandable.

```
Data Display
K1        0.400000      #this is p̄
K2        -1.22474      #this is Zsample
```

Compute the p-value

f.   The p-value is the tail area from the sample Z value. Here the sample Z value (the sample p_s value converted to a Z) was -1.22474. The tail area for z = -1.22474 is F(-1.22474). Type
```
MTB > cdf -1.22474;
SUBC> normal mu 0 stdev 1 .
```

g.   The result is shown below. The tail area for the sample z of -1.2247 is 0.1103. Since this was a two-tailed test, the p-value is twice this tail area, so the p-value is 0.2206 . Since this represents a large risk of a Type I error if the null hypothesis is rejected, we were right to not reject H_0 in the previous problem. The p-value is much higher than a typical α level of risk of 0.05. Since the risk of an error when rejecting H_0 is 0.2206 > 0.05, it is reasonable to not reject H_0. (Note that the text result may differ slightly when using the table on rounded values.)
```
Cumulative Distribution Function
Normal with mean = 0 and standard deviation = 1.00000
    x       P( X <= x)
-1.2247      0.1103
```

χ^2 Test for Differences in Two Proportions (Independent Samples)

One advantage of using the Chi-Square test rather than the Z test is that it can be extended to more than 2 samples. In this situation it can only be a two-tailed test.

Here the goal is to use the two sample proportions P_{s1} and P_{s2} to test some hypothesis regarding the difference in two population proportions p_1 and p_2. Then the test statistic is approximately

$$\chi^2 = \sum_{\text{all cells}} \frac{(f_o - f_e)^2}{f_e} ,$$

where f_o is the *observed frequency* in a particular cell, and f_e is the *expected frequency* in that cell if the null hypothesis is true.

PROBLEM 15.23

This uses the same data as 15.13. Is there a difference in the proportion of accountants who found Company A's reports *understandable* versus the proportion of accountants who found Company B's reports *understandable*? In a sample of 50 accountants, 17 found Company A's reports understandable. In a separate sample of 50 accountants, 23 found Company B's reports understandable. Test the hypotheses at $\alpha = .10$ level of significance.

a.   The hypotheses are H_0: $p_1 = p_2$ versus H_1: $p_1 \neq p_2$.

b.   Let Minitab determine the correct χ^2 value (which could be obtained from a table.) For this type of test there is only an upper critical value, which is F(.90). There is (r-1)(c-1) = (2-1)(2-1) = 1 degree of freedom. At the Session Window type
```
MTB > invcdf .90;
SUBC> chisquare 1 df.
```

c.   The display is shown below. The upper critical χ^2 value is 2.7055, which is the test value.

```
Inverse Cumulative Distribution Function
Chisquare with 1 d.f.
P( X <= x)            x
   0.9000         2.7055
```

d.   Now to decide whether the sample provides sufficient support to reject H_0, we compare the test χ^2 value with the sample χ^2 value. If the sample result converted to a χ^2 is greater than the critical $\chi^2 = 2.7055$, this will be sufficient evidence to reject H_0 in favor H_1. Minitab provides a simple way to determine the sample χ^2 value.

In column c1 (or whichever column you choose) type the heading **CO-A** and in column c2 type the heading *CO-B*. In the first two rows of *CO-A* enter 17 and 33 (since there were 50 total reading the reports for *CO-A* and 17 found them good and 33 didn't). Similarly, in column *CO-B* enter 23 and 27.

e.   Select **Stat>Tables>Chisquare Test**, and for *Columns containing the table* double-click on *CO-A* and then on *CO-B*. Click *OK*.

f.   If you typed rather than using the menus,
```
MTB > ChiSquare  'CO-A' 'CO-B'.
```

g.   The result is shown below. Note that the observed values are shown, as well as the expected values assuming equal proportions for the two populations. Since the sample χ^2 value of 1.50 is less than the critical χ^2 value of 2.7055, then one is lead by this sample result to not reject H_0. Therefore the conclusion is that there is no difference in the proportions which thought the two annual reports were understandable; this is the same conclusion reached when performing the test using the Z approximation.

```
Chi-Square Test
Expected counts are printed below observed counts
            CO-A     CO-B    Total
    1        17       23       40
          20.00    20.00

    2        33       27       60
          30.00    30.00

Total        50       50      100

ChiSq =  0.450 +  0.450 +
         0.300 +  0.300 = 1.500
df = 1, p = 0.221
```

p-value

h.   The text doesn't ask us to compute the p-value, but it is automatically provided. The p-value is the tail area from the sample χ^2 value. The p-value is shown on the output as 0.221. Since this represents a large risk of a Type I error if the null hypothesis is rejected, we were right to not reject H_0. The p-value is much higher than a typical α level of risk of 0.05. Since the risk of an error when rejecting H_0 is $0.221 > 0.05$, it is reasonable to not reject H_0. More specifically, the p-value of 0.221 is higher than the specified α level of 0.10 .

   i.   The graph is shown for this problem. Note that the shape for the χ^2 density function when there is only 1 degree of freedom is different from the usual shape shown for a Chi-Square problem.

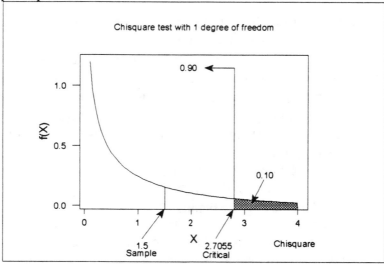

χ^2 Test for Differences in c Proportions (Independent Samples)

PROBLEM 15.30

Is there a difference in attitude toward the trimester among undergraduate students, graduate students, and faculty? Test this at $\alpha = 0.01$ level of significance.

   a.   The hypotheses are H_0: $p_1 = p_2 = p_3$ versus H_1: at least one $p_i \neq p_j$.

   b.   Let Minitab determine the correct χ^2 value (which could be obtained from a table.) For this type of test there is only an upper critical value, which is F(.99). There are $(c-1) = (3-1) = 2$ degrees of freedom. At the Session Window type

```
MTB > invcdf .99;
SUBC> chisquare 2 df.
```

   c   The display is shown below. The upper critical χ^2 value is 9.2103, which is the test value.

```
Inverse Cumulative Distribution Function
Chisquare with 2 d.f.
P( X <= x)            x
   0.9900        9.2103
```

   d.   Now to decide whether the sample provides sufficient support to reject H_0, we compare the test χ^2 value with the sample χ^2 value. If the sample result converted to a χ^2 is greater than the critical $\chi^2 = 9.2103$, this will be sufficient evidence to reject H_0 in favor H_1. Minitab provides a simple way to determine the sample χ^2 value.
        In column c1 (or whichever column you choose) type the heading **UGrad** and in column c2 type the heading **Grad**, and in c3 type **Faculty**. In the first two rows of

UGrad enter **63** and **37**. Similarly, in column *Grad* enter **27** and **23**, and in column *Faculty* enter **30** and **20**.

e.   Select **Stat>Tables>Chisquare Test**, and for *Columns containing the table* double-click on *UGrad* and then on *Grad* and then on *Faculty*. Click *OK*.

f.   If you typed rather than using the menus,
```
MTB > ChiSquare  'UGrad' 'Grad' 'Faculty'.
```

g.   The result is shown below. Note that the observed values are shown, as well as the expected values assuming equal proportions for the two populations. Since the sample χ^2 value of 1.125 is less than the critical χ^2 value of 9.2103, then one is lead by this sample result to not reject H_0. Therefore the conclusion is that there is no difference in the attitudes of the three groups toward trimesters.

```
Chi-Square Test
Expected counts are printed below observed counts
            UGrad      Grad   Faculty     Total
    1          63        27        30       120
            60.00     30.00     30.00

    2          37        23        20        80
            40.00     20.00     20.00

Total         100        50        50       200

ChiSq =   0.150 +   0.300 +   0.000 +
          0.225 +   0.450 +   0.000 = 1.125
df = 2, p = 0.570
```

p-value

h.   The text doesn't ask us to compute the p-value, but it is automatically provided. The p-value is the tail area from the sample χ^2 value. The p-value is shown on the output as 0.570. Since this represents a large risk of a Type I error if the null hypothesis is rejected, we were right to not reject H_0. The p-value is much higher than a typical α level of risk of 0.05. Since the risk of an error when rejecting H_0 is 0.570 > 0.05, it is reasonable to not reject H_0. More specifically, this p-value of 0.570 is much higher than the specified α level of 0.01.

χ^2 Test for Independence

The goal is to test whether there is some relationship between two categorical variables.

PROBLEM 15.40

Is there evidence of a relationship between commuting time and stress? Test this at $\alpha=.01$ level of significance.

a. H_0: The factors commuting time and stress are independent
         H_1: they are not independent .

b. Let Minitab determine the correct χ^2 value (which could be obtained from a table.) For this type of test there is only an upper critical value, which is F(.99). There are $(r-1)(c-1) = (3-1)(3-1) = 4$ degrees of freedom. At the Session Window type
```
MTB > invcdf .99;
SUBC> chisquare 4 df.
```
c. The display is shown below. The upper critical χ^2 value is 13.2767, which is the test value.
```
Inverse Cumulative Distribution Function
Chisquare with 2 d.f.
P( X <= x)        x
   0.9900      13.2767
```
d. Now to decide whether the sample provides sufficient support to reject H_0, we compare the test χ^2 value with the sample χ^2 value. If the sample result converted to a χ^2 is greater than the critical $\chi^2 = 13.2767$, this will be sufficient evidence to reject H_0 in favor H_1. Minitab provides a simple way to determine the sample χ^2 value.

e. Type the headings for the different levels of stress: In column c1 (or whichever column you choose) type the heading **High** and in column c2 type the heading **Moder**, and in c3 type **Low**. In the first three rows of *High* enter **9**, **17** and **18**. Similarly, in column *Moder* enter **5**, **8**, and **6**, and in column *Low* enter **18**, **28**, and **7**.

f. Select **Stat>Tables>Chisquare Test**, and for *Columns containing the table* double-click on *High* and then on *Moder* and then on *Low*. Click *OK*.

g. If you typed rather than using the menus,
```
MTB > ChiSquare  'High' 'Moder' 'Low'.
```

h. The result is shown below. Note that the observed values are shown, as well as the expected values assuming independence. Since the sample χ^2 value of 9.831 is less than the critical χ^2 value of 13.2767, then one is lead by this sample result to not reject H_0. Therefore the conclusion is that there is no relationship between commuting time and stress: the two factors are independent.

```
Chi-Square Test
Expected counts are printed below observed counts

            High    Moder    Low    Total
    1          9        5      18       32
           12.14     5.24   14.62

    2         17        8      28       53
           20.10     8.68   24.22

    3         18        6       7       31
           11.76     5.08   14.16

Total         44       19      53      116

ChiSq =  0.811 +  0.011 +  0.781 +
         0.479 +  0.053 +  0.591 +
         3.313 +  0.168 +  3.623 = 9.831
df = 4, p = 0.044
```

p-value
i. Compute the p-value. The p-value is the tail area from the sample χ^2 value. The p-value is shown on the output as 0.044. Since this represents a small risk of a Type I error if the null hypothesis is rejected, we might ordinarily be led to reject H_0. However, since for this example the level of significance was 0.01, then we are led to not reject H_0, since the p-value is higher than the specified α level of risk of 0.01 .

McNemar Test for Differences in Two Related Proportions

This test may be used to determine whether there is a difference between two related proportions (a two-tailed test), or to determine whether one group has a higher proportion than the other (a one-tailed test). This may be viewed as a repeated-response experiment, since typically there are two responses from the same source.

When the data are placed into a contingency table, where the responses pertain to a *before* and *after* position, then the value of Z may be approximated by

$$Z \sim \frac{B - C}{\sqrt{B + C}}.$$

	After: Case 1	After: Case 2
Before: Case 1	A	B
Before: Case 2	C	D

PROBLEM 15.57

Is there a difference in the proportion of voters who favor Candidate A prior to and after a debate? The data can be summarized as shown below. Test this at $\alpha = .01$ level of significance.

	After: Candidate A	After: Candidate B
Before: Candidate A	269	21
Before: Candidate B	36	174

a. $H_0: p_1 = p_2$ versus $H_1: p_1 \neq p_2$.

b.   Let Minitab determine the correct Z values (which could be obtained from a table.) The two critical values are for $F(0.005)$ and $F(0.995)$. The upper critical Z is for $F(0.995)$. At the Session Window type
```
MTB > invcdf .995;
SUBC> normal mu 0 stdev 1.
```

c.   The display is shown below. The upper critical Z value is 2.5758, and because of symmetry the lower critical Z value is -2.5758, which are the test values.
```
Inverse Cumulative Distribution Function
Normal with mean = 0 and standard deviation = 1
P( X <= x)          x
0.9950          2.5758
```

d.   Now to decide whether the sample provides sufficient support to reject H_0, we compare the test Z-value with the sample Z-value. If the sample result converted to a Z-score is less than -2.5758 or more than 2.5758, this will be sufficient evidence to reject H_0 in favor H_1. Let Minitab provide the computation: type
```
MTB > let k1 = (21 - 36)/ sqrt(21 + 36)
MTB > print k1
```

e.   The result is shown below. Since the sample Z value of -1.98680 is greater than the critical Z value of -2.5758 (it is inside the acceptance region), then one is lead by this sample result to not reject H_0. Therefore the conclusion is that there is no difference in the proportions who favored candidate A prior to and after the debate.
```
Data Display
K1          -1.98680
```

p-value

    f.   The p-value is the tail area from the sample Z value. Here the sample Z value was -1.98680. The tail area for $z = -1.98680$ is $F(-1.98680)$. Type

```
MTB > cdf -1.98680;
SUBC> normal mu 0 stdev 1 .
```

    g.   The result is shown below. The tail area for the sample Z of -1.9868 is 0.0235. Since this was a two-tailed test, the p-value is twice this tail area, so the p-value is 0.047 . Since this represents a small risk of a Type I error if the null hypothesis is rejected, we might ordinarily be led to reject H_0. However, since for this example the level of significance was 0.01, then we are led to not reject H_0, since the p-value is higher than the specified α level of risk of 0.01 . Since the risk of an error when rejecting H_0 is $0.047 > 0.01$, it is reasonable to not reject H_0.

```
Cumulative Distribution Function
Normal with mean = 0 and standard deviation = 1.00000
    x       P( X <= x)
 -1.9868     0.0235
```

STATISTICAL APPLICATIONS IN QUALITY AND PRODUCTIVITY MANAGEMENT

This chapter provides an introduction to the history of quality, and demonstrates the use of a number of control charts.

The focus for this chapter is statistical applications in quality and productivity management.

1. Charts provide a simple and useful method for managers to evaluate processes. There are numerous charts which may be used, and each chart has a different story to tell, so more than one should usually be employed. For instance, control of *variability* of a process is usually of paramount importance (since the variability affects the mean, range, and other measures), so typically a first look at a process would include one or more charts to evaluate variability, such as the R (Range) and/or the s (standard deviation) charts.

2. These control charts are essentially equivalent to hypothesis testing. The hypotheses typically are of the form H_o: The process is in control versus H_1: The process is not in control. The critical values are usually the UCL (Upper Control Limit) and the LCL (Lower Control Limit). These Control Limits are typically computed by something equivalent to *statistic ± 3 standard deviations*.

3.   The charts which we'll discuss are:

Chart	Purpose
Fishbone	Represent Causes and Effects
Control Charts in general	Typically a chart with center line and upper and lower Control Limits.
$\bar{x}$-chart	Monitor the mean.
R-chart	Monitor the variability, using the range.
s-chart	Monitor the variability, using the standard deviation.
p-chart	Monitor the proportion of defectives.
np-chart	Monitor the number of defectives.
c-chart	Monitor the number of defects per item.
u-chart	Monitor the number of defects per item when subgroup sizes are different.
X-chart	Monitor the mean, using the individual observations.

Fishbone

The Fishbone diagram is often referred to as a Cause-and-Effect diagram, because its purpose is to represent causes and effects.

The example to be presented is a *Minitab* data file.

    a.   Open the file EXH_QC.MTW by using the menu **File>Open Worksheet** and typing the file name *exh_qc.mtw* and its location, which may require a path similar to \mtbwin\data.

    b.   Select **Stat>SPC>Cause-and-Effect...**

    c.   You are presented with a dialog box with one set of boxes (under the heading *Causes*) which must be filled out, and another set of boxes (under the heading *Label*) which have already been filled out for this example.

        1.   For each of the blank boxes under the heading *Causes*, click in the box and then double-click on the *Cause* in the list at the left. What you will double-click is shown below in **bold**.

Branch	Causes *Type these*	Label *These already exist*
1	Man	Men
2	Machine	Machine
3	Material	Materials
4	Method	Methods
5	Measure	Measurements
6	Enviro	Environment

        2.   In the box labeled *Effect* type `Surface Flaws`

        3.   In the box labeled *Title* type `Cause and Effect Diagram`

        4.   Click on *OK*.

d.   The resulting diagram is shown below.

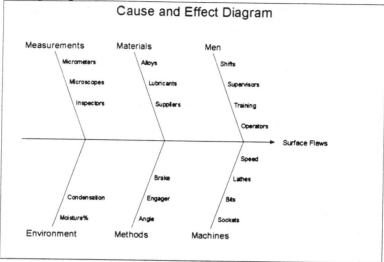

e.   If you typed rather than using the menus,

```
MTB >  %Fishbone;
SUBC>    Br1 'Man';
SUBC>    Br2 'Machine';
SUBC>    Br3 'Material';
SUBC>    Br4 'Method';
SUBC>    Br5 'Measure';
SUBC>    Br6 'Enviro';
SUBC>    Nam1 'Men';
SUBC>    Nam2 'Machines';
SUBC>    Nam3 'Materials';
SUBC>    Nam4 'Methods';
SUBC>    Nam5 'Measurements';
SUBC>    Nam6 'Environment';
SUBC>    Effect 'Surface Flaws';
SUBC>    Title 'Cause and Effect Diagram'.
```

The np Chart

This is used when the subgroups are all of the same size, as an alternative to the p chart. The control limits may be computed by

$$\overline{X} \pm 3\sqrt{\overline{X}(1-\overline{p})}, \text{ where}$$

$$\overline{p} = \frac{\sum_{i=1}^{k} X_i}{nk}.$$

PROBLEM 16.7

The number of foul shots made by a player has been recorded. Is the number in statistical control?

    a.   If the data aren't current, open the file by using the menu **File>Other Files>Import ASCII Data**; type **c1-c2** for *column* (or whatever columns you choose) and click *OK*. Give the file name *foulspc.dat* and its location. Type the heading **Made** for column c1 and **Tries** for column c2.

    b.   Select **Stat>Control Charts>NP...**

    c.   For *Variable* double-click on *Made* column, and for *Subgroups in* click in the box then double-click on *Tries*.

    d.   Axis tick marks: Click on **Frame>Tick** and you will see a screen with the following headings. The numbers enclosed in a box are the ones to enter. In Row 1, enter **5** for the Number of Major ticks and enter **9** for the number of Minor Ticks. Then Click OK.

	Direction	Side	Positions	Number of Major	Number of Minor
1	X			5	9

    e.   Click on *OK* twice to obtain the graph.

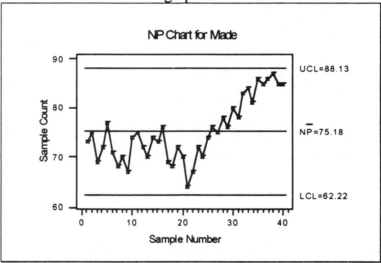

    f.   If you typed rather than used the menus,

```
MTB > NPChart 'Made' 'Tries';
SUBC>    Tick 1;
SUBC>      Number 5 9;
SUBC>    Tick 2;
SUBC>    Symbol;
SUBC>    Connect.
```

The p Chart

The control limits for proportions are shown below:

$$\bar{p} = \frac{\sum_{i=1}^{k} \hat{p}_i}{k \text{ samples}} \Rightarrow \text{the center line is } \bar{p}, \text{ where } \hat{p} \text{ is the individual sample proportion.}$$

$$LCL = \bar{p} - (3)\sqrt{\frac{\bar{p}(1-\bar{p})}{n}} \qquad \text{and}$$

$$UCL = \bar{p} + (3)\sqrt{\frac{\bar{p}(1-\bar{p})}{n}}.$$

PROBLEM 16.12

Over a period of 30 days there are data values on the number of orders for that day and the number of corrections which were required. Is the process out of control?

a. If the data aren't current, open the file by using the menu **File>Other Files>Import ASCII Data**; type **c1-c2** for *column* (or whatever columns you choose) and click *OK*. Give the file name *telespc.dat* and its location. Type the heading **Orders** for column c1 and **Correct** for column c2.

b. Select **Stat>Control Charts>p...**

c. For *Variable* double-click on *Correct* column, and for *Subgroups in* double-click on *Orders*.

d. Title: Click on **Annotation>Title** and type **p-Chart for Ex 16.12**

e. Click in the first box under *Text Size* and type **1** to reduce the title size. Click *OK*.

f. Axis tick marks: Click on **Frame>Tick** and you will see a screen with the following headings. The numbers enclosed in a box are the ones to enter. In Row 1, enter **4** for the Number of Major ticks and enter **9** for the number of Minor Ticks. Then Click OK.

	Direction	Side	Positions	Number of Major	Number of Minor
1	X			4	9

g. Click on *OK* twice to obtain the graph.

h.   Note that the value for $\bar{p} = .1091$, and the UCL and LCL are also shown. Note further that observation 9 is nearly out of its UCL; observations 26 and 30 are out of limits. These should be analyzed to see if the process needs to be adjusted.

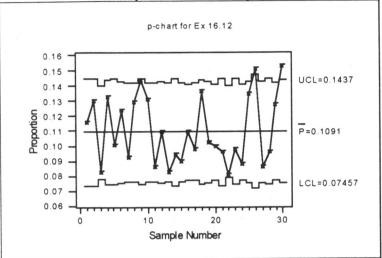

i.   If you typed rather than used the menus,

```
MTB > PChart 'Correct' 'Orders';
SUBC>    Title "p-chart for Ex 16.12";
SUBC>      TSize 1;
SUBC>    Tick 1;
SUBC>      Number 4 9;
SUBC>    Tick 2;
SUBC>    Symbol;
SUBC>    Connect.
```

The c Chart: A Control Chart for the Number of Occurrences per Unit.

This chart monitors the *number of defects or imperfections per item or unit* (such as blemishes per yard of cloth). A p-chart would be appropriate when both the defectives and nondefectives can be counted, such as inspecting parts. The c-chart is appropriate when you can count the number of things which did occur, but not the number which didn't occur. For instance, you can count the number of accidents during a period of time, but *not* the number which didn't occur. You can count the number of scratches on a polished surface, the number of bacteria present in a water sample, and the number of crimes committed during the month of August, but not the ones which didn't. The Poisson distribution is used to model many random events (such as calls into a telephone exchange, or cars passing a toll booth), and it provides the basis for the c-chart. For large n, the Poisson can be approximated by the normal distribution.

1.   The form of the interval is similar to those we've used previously:

$$\bar{c} = \frac{\sum_{i=1}^{n} c_i}{n \text{ in the sample}} \Rightarrow \text{the center line is } \bar{c}.$$

2.   Compute the critical values. (Since we assume the defects per item follow a Poisson distribution, then we use the fact that for the Poisson distribution, the mean $\bar{c}$ equals the variance, also $\bar{c}$).

$$\text{LCL} = \bar{c} - (3)\sqrt{\bar{c}} \qquad \text{and}$$

$$\text{UCL} = \bar{c} + (3)\sqrt{\bar{c}} \; .$$

PROBLEM 16.18

Over a period of 30 days there are data values on the number of orders for that day and the number of corrections which were required. Is the process out of control?

a.   If the data aren't current, open the file by using the menu **File>Other Files>Import ASCII Data**; type **c1** for *column* (or whatever column you choose) and click *OK*. Give the file name *baggage.dat* and its location. Type the heading **Claims** for your selected column.

b.   Select **Stat>Control Charts>c...**

c.   For *Variable* double-click on the *Claims* column.

d.   Title: Click on **Annotation>Title** and type **c-Chart for Ex 16.18**
Click in the first box under *Text Size* and type **1** to reduce the title size. Click *OK*.

e.   Axis tick marks: Click on **Frame>Tick** and you will see a screen with the following headings. The numbers enclosed in a box are the ones to enter. In Row 1, enter **4** for the Number of Major ticks and enter **9** for the number of Minor Ticks. Then Click OK.

	Direction	Side	Positions	Number of Major	Number of Minor
1	X			4	9

f.   Click on *OK* twice to obtain the graph.

g.   Note that the value for $\bar{c}$ = 26.87, and the UCL and LCL are also shown. Note further that observation 14 exceeds its UCL. This observation should be analyzed to see if the process needs to be adjusted.

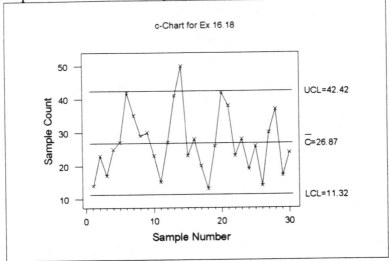

h.   If you typed rather than used the menus,

```
MTB > CChart Claims;
SUBC>    Title "c-Chart for Ex 16.18";
SUBC>     TSize 1;
SUBC>    Tick 1;
SUBC>     Number 4 9;
SUBC>    Tick 2;
SUBC>    Symbol;
SUBC>    Connect.
```

The R Chart: A Control Chart for Dispersion

The R-chart has the same form as the $\bar{x}$-chart: a center line and symmetric upper and lower control limits (even though the distribution of the sample range is not normal). The strategy is to consider the process out of control if one observed sample range (from one sample) is outside the computed interval.

The value for $\bar{R}$ is computed like $\bar{X}$, but it is now the average of the sample **ranges**,

$$\bar{R} = \frac{\sum_{i=1}^{k} R_i}{k} \ .$$

The form of the interval is similar to the $\bar{X}$-chart:

$$\bar{R} \pm 3\,\bar{R}\,\frac{d_3}{d_2}$$ gives the upper and lower limits, where the constants d_2 and d_3 are in table E.13 in the text.

PROBLEM 16.21
The data are a group of waiting times at a local bank. Is the process out of control?

a.   If the data aren't current, open the file by using the menu **File>Other Files>Import ASCII Data**; type **c1-c2** for *column* (or whatever columns you choose) and click *OK*. Give the file name *banktime.dat* and its location. Type the heading **Time** for c1 and **Day** for c2.

b.   Select **Stat>Control Charts>R...**

c.   For *Variable* double-click on the *Time* column.

d.   For *Subgroups in* click in the box and then double-click on *Day*.

e.   Axis tick marks: Click on **Frame>Tick** and you will see a screen with the following headings. The numbers enclosed in a box are the ones to enter. In Row 1, enter **3** for the Number of Major ticks and enter **9** for the number of Minor Ticks. Then Click OK.

	Direction	Side	Positions	Number of Major	Number of Minor
1	X			3	9

f.   Click on *OK* twice to obtain the graph.

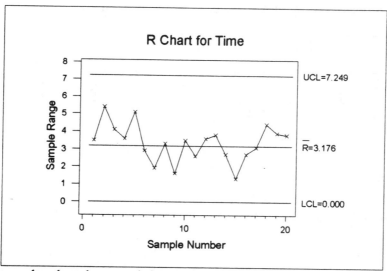

g.   If you typed rather than used the menus,

```
MTB > RChart 'Time' 'Day';
SUBC>    Tick 1;
SUBC>      Number 3 9;
SUBC>    Tick 2;
SUBC>    Symbol;
SUBC>    Connect.
```

The $\overline{X}$ Chart

$$\overline{\overline{X}} = \frac{\sum_{i=1}^{k} \overline{X}_i}{k} \text{ and } \overline{R} = \frac{\sum_{i=1}^{k} R_i}{k}, \text{ so that the control limits are } \overline{\overline{X}} \pm 3\frac{\overline{R}}{d_2\sqrt{n}}, \text{ and the constants } d_2$$

and d_3 are in table E.13 in the text.

PROBLEM 16.21
The data are a group of waiting times at a local bank. Is the process out of control?

a.   If the data aren't current, open the file by using the menu **File>Other Files>Import ASCII Data**; type **c1-c2** for *column* (or whatever columns you choose) and click *OK*. Give the file name *banktime.dat* and its location. Type the heading **Time** for c1 and **Day** for c2.

b.   Select **Stat>Control Charts>Xbar...**

c.   For *Variable* double-click on *Time* column.

d.   For *Subgroups in* click in the box and then double-click on *Day*.

e.   Axis tick marks: Click on **Frame>Tick** and you will see a screen with the following headings. The numbers enclosed in a box are the ones to enter. In Row 1, enter **3** for the Number of Major ticks and enter **9** for the number of Minor Ticks. Then Click OK.

	Direction	Side	Positions	Number of Major	Number of Minor
1	X			3	9

f.   Click on *OK* twice to obtain the graph.

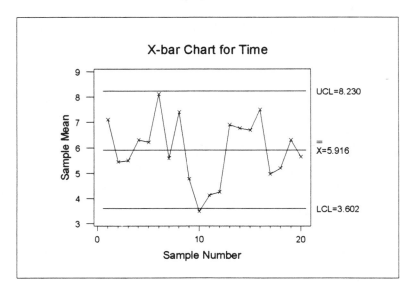

g. If you typed rather than used the menus,

```
MTB > XbarChart 'Time' 'Day';
SUBC>    Tick 1;
SUBC>     Number 3 9;
SUBC>    Tick 2;
SUBC>    Symbol;
SUBC>    Connect.
```

Combined $\overline{X}$ Chart and the R-Chart (not a separate heading in the text):

We may wish to use the $\overline{X}$ Chart and the R-Chart together. The $\overline{X}$ Charts are useful for telling when a process has gone out of control in this sense: when the ***mean of the sample exceeds a control limit.*** The assumption is that the ***variance of the process*** is constant, and that if something goes wrong with the process it will only affect the mean. If the process begins to go wrong by increasing the ***variance*** of the parts, the $\overline{X}$ Chart will not catch this change. The R-Chart or Range-Chart will not catch changes in the means like the $\overline{X}$-Chart will, but the R-Chart ***will*** catch changes in variability. Minitab provides the ability to produce the two graphs together.

1. To obtain the $\overline{X}$-chart and the R-chart together,
    a. Select **Stat>Control Charts>Xbar and R...**
    b. For *Variable* double-click on *Time* column.
    c. For *Subgroups in* click in the box and then double-click on *Day.*
    d. Click in the box for the *Title* and type **X-Bar and R charts**
    e. Click *OK*.

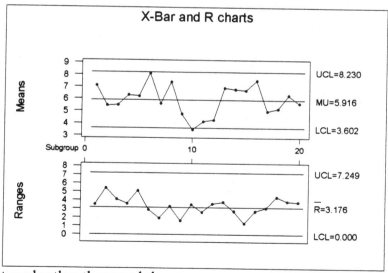

f. If you typed rather than used the menus,

```
MTB > %XRChart 'Time' 'Day';
SUBC>    Title 'X-Bar and R charts'.
```

Control Charts for Individual Values (X Charts)

For those occasions when it is not feasible to obtain a large sample size, we may treat each observation as its own subgroup. The *moving range* is used as an estimate of the variation, where the moving range is

$$MR_i = |X_{i-1} - X_i|$$

$$\text{and } \overline{MR} = \frac{\sum_{i=1}^{k} MR_i}{k-1} .$$

Then the control limits are $\overline{\overline{X}} \pm 3 \dfrac{\overline{MR}}{d_2}$.

PROBLEM 16.26
A machine produces packages of salt with a desired weight. Is the process out of control?

1. If the data aren't current, open the file by using the menu **File>Other Files>Import ASCII Data**; type **c1-c2** for *column* (or whatever columns you choose) and click *OK*. Give the file name *salt.dat* and its location. Type the heading **Package** for c1 and **Weight** for c2.

Note that in Minitab you may wish to obtain a chart of moving ranges. For completeness, this chart is described.

2. To obtain the moving range chart
   a. Select **Stat>Control Charts>Moving Range...**
   b. For *Variable* double-click on *Weight* column.
   c. Leave the *Moving range span* setting at 2.
   d. Axis tick marks: Click on **Frame>Tick** and you will see a screen with the following headings. The numbers enclosed in a box are the ones to enter. In Row 1, enter **6** for the Number of Major ticks and enter **9** for the number of Minor Ticks. Then Click OK.

	Direction	Side	Positions	Number of Major	Number of Minor
1	X			6	9

e.   Click *OK* twice.

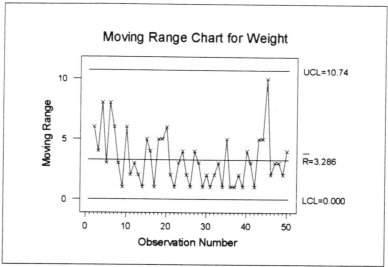

f.   If you typed rather than used the menus,

```
MTB > MRChart 'Weight';
SUBC>    RSpan 2;
SUBC>    Tick 1;
SUBC>      Number 6 9;
SUBC>    Tick 2;
SUBC>    Symbol;
SUBC>    Connect.
```

The chart we are really looking for is the individual chart X where the moving range provides the estimate of variation.

3.   To obtain the individual chart,

    a.   **Select Stat>Control Charts>Individuals...**

    b.   For *Variable* double-click on the *Weight* column.

    c.   Leave the *Moving range span* setting at 2.

    d.   Axis tick marks: Click on **Frame>Tick** and you will see a screen with the following headings. The numbers enclosed in a box are the ones to enter. In Row 1, enter **6** for the Number of Major ticks and enter **9** for the number of Minor Ticks. Then Click OK.

	Direction	Side	Positions	Number of Major	Number of Minor
1	X			6	9

e.   Click *OK* twice.

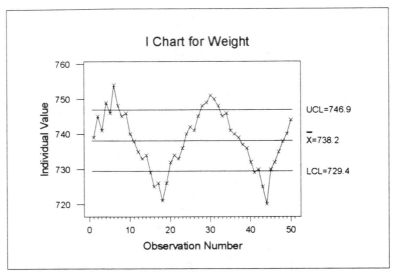

f.   If you typed rather than used the menus,
```
MTB > IChart 'Weight';
SUBC>    RSpan 2;
SUBC>    Tick 1;
SUBC>      Number 6 9;
SUBC>    Tick 2;
SUBC>    Symbol;
SUBC>    Connect.
```

Minitab provides the two charts together.
4.   To obtain the Individual chart and Moving Range chart together,
    a.   Select **Stat>Control Charts>I-MR...**
    b.   For *Variable* double-click on *Weight* column.
    c.   Leave the *Moving range span* setting at 2.
    d.   Click on the box for the title and type **Individual and Moving Range**

e.   Click *OK*.

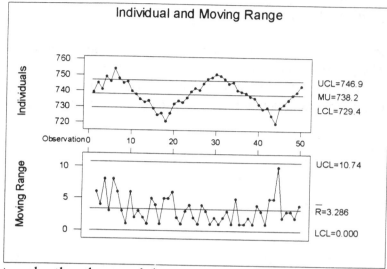

f.   If you typed rather than used the menus,

```
MTB > %IMRChart 'Weight';
SUBC>    RSpan 2;
SUBC>    Title 'Individual and Moving Range'.
```

[Not in Text:]

There are several more charts which may be of interest, but which are not described in the text. For completeness they are presented here: the u-chart and the s-chart.

The u-Chart: A Control Chart for the Number of Occurrences per Unit with differing sizes of subgroups. The c-chart monitors the *number of defects or imperfections per item or unit*, when the subgroups are the same size. When they are not the same size, the u-chart is the appropriate chart.

The form of the interval is similar to those we've used previously:

$$\bar{u} = \frac{\sum_{i=1}^{k} c_i}{\sum_{i=1}^{k} n_i} \Rightarrow \text{the center line is } \bar{u}.$$

The control limits *for each day or unit* are therefore

$$LCL = \bar{u} - (3)\sqrt{\frac{\bar{u}}{n_i}} \qquad \text{and}$$

$$UCL = \bar{u} + (3)\sqrt{\frac{\bar{u}}{n_i}}.$$

This PROBLEM is not in the text. It is problem 7.20 in the text *Business Statistics for Quality and Productivity* by Levine, Ramsey, and Berenson.

Over 31 days, a bookkeeper has differing numbers of transaction per day, and differing numbers of errors per day. Therefore, for each day there are upper and lower control limits for the number of errors per day. A c-chart would be appropriate for this situation if the bookkeeper had the same number of transactions each day, but here they differ. Is the process out of control?

a. The data are as shown below. (If you entered the data into Minitab you would not need the *Day* information.)

Day	Number of Transactions per day	Number of Errors per day
1	21	4
2	33	3
3	24	4
4	28	5
5	22	5
6	50	2
7	25	1
8	25	3
9	26	2
10	21	1
11	23	1
12	29	3
13	34	6
14	24	4
15	28	0
16	31	8
17	32	1
18	67	7
19	22	1
20	36	2
21	22	2
22	52	3
23	25	0
24	22	1
25	23	3
26	39	2
27	22	3
28	21	4
29	33	4
30	32	6
31	46	2

b. The headings are **Transact** and **Errors**.
c. Select **Stat>Control Charts>u...**
d. For *Variable* double-click on *Errors* column.
e. For *Subgroups in* click in the box and then double-click on *Transact* .
f. Axis tick marks: Click on **Frame>Tick** and you will see a screen with the following headings. The numbers enclosed in a box are the ones to enter. In Row 1, enter **4** for

the Number of Major ticks and enter **9** for the number of Minor Ticks. Then Click OK.

	Direction	Side	Positions	Number of Major	Number of Minor
1	X			4	9

g.   Click on *OK* twice to obtain the graph.

h.   Note that the value for $\bar{u} = 0.099$, and the UCL and LCL are also shown. (You may wish to multiply the center point and the limits by 100.) There is only one observation which is nearly too large, but mostly the process is in control.

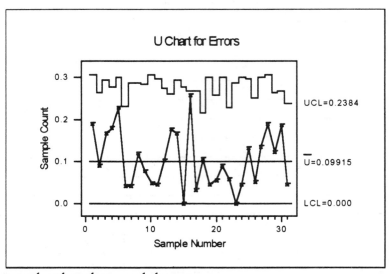

i.   If you typed rather than used the menus,

```
MTB > UChart 'Errors' 'Transact';
SUBC>    Tick 1;
SUBC>      Number 4 9;
SUBC>    Tick 2;
SUBC>    Symbol;
SUBC>    Connect.
```

The s-Chart

The standard deviation chart provides another mechanism to monitor the variability of some process. Like the R-chart, the distribution of the standard deviations is not normal (nor is it symmetric). However, we still will determine symmetric upper and lower critical values for the interval.

1.   The form of the interval is similar to the R-chart:

        $\bar{s}$ ± some sampling error, and the limits will be estimated.

2.   The steps for constructing an s-chart are as follows.
    a.   Compute the sample standard deviation for each of the k samples:

$$\text{the } i^{\text{th}} \text{ such standard deviation} = s_i = \sqrt{\frac{\sum\limits_{j=1}^{n}(x_{ij} - \bar{x}_i)^2}{n-1}} = \sqrt{\frac{\Sigma(x^2) - \frac{(\Sigma x)^2}{n}}{n-1}}.$$

    b.   Compute the grand mean of the standard deviations: $\bar{s} = \dfrac{\sum\limits_{i=1}^{k} s_i}{k}$.

    c.   Find in a suitable table the value of the constant B_3, and the value of B_4, for the appropriate sample size.
    d.   Compute the estimates of the limits:
        $LCL = B_3 \bar{s}$ and $UCL = B_4 \bar{s}$.

PROBLEM 16.21 revisited

The data are a group of waiting times at a local bank. Is the process out of control?

    a.   If the data aren't current, open the file by using the menu **File>Other Files>Import ASCII Data**; type **c1-c2** for *column* (or whatever columns you choose) and click *OK*. Give the file name *banktime.dat* and its location. Type the heading **Time** for c1 and **Day** for c2.
    b.   Select **Stat>Control Charts>S...**
    c.   For *Variable* double-click on *Time* column.
    d.   Click *Subgroups in*, then click in the box and then double-click on *Day*.
    e.   **Axis tick marks**: Click on **Frame>Tick** and you will see a screen with the following headings. The numbers enclosed in a box are the ones to enter. In Row 1, enter **3** for the Number of Major ticks and enter **9** for the number of Minor Ticks. Then Click OK.

	Direction	Side	Positions	Number of Major	Number of Minor
1	X			3	9

    f.   Click on *OK* twice to obtain the graph.

g.   The process seems well in control with respect to variability.

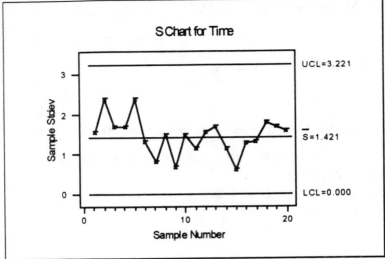

h.   If you typed rather than used the menus,

```
MTB > SChart 'Time' 'Day';
SUBC>    Tick 1;
SUBC>      Number 3 9;
SUBC>    Tick 2;
SUBC>    Symbol;
SUBC>    Connect.
```

SIMPLE LINEAR REGRESSION AND CORRELATION

This chapter develops the simple linear regression model as a means of using one variable to predict another. The correlation model is also presented.

The Scatter Diagram
This is often the initial step in analyzing data

PROBLEM 17.2
The data represent for 14 stores the square footage of the store and the annual sales. Initially, we just want to get a visual representation of the data.

a. If the data aren't current, open the file by using the menu **File>Other Files>Import ASCII Data**; type **c1-c3** for *column* (or whatever columns you choose) and click *OK*. Give the file name *site.dat* and its location. Type the heading **Store** for c1, **SqFt** for c2, and **Sales** for c3.
b. Select **Graph>Plot**
c. For *Graph Variables*: for *Y* double-click on *Sales*; for X double-click on *SqFt*.
d. For *Display*, choose *Symbol*, and for *For each*, choose *Graph*.
e. Click on *Symbol* and click on the *Edit Attributes* button and under *Type* click on the arrow and select *Solid circle*, and click *OK*.
f. Title: Click on **Annotation>Title** and type **Scatter Diagram**
    Click in the first box under *Text Size* and type **1** to reduce the title size. Click *OK*. This will decrease the size of the *Title* font size.
g. Click on *OK* to obtain the graph.

161

h.   The relationship appears to be fairly linear.

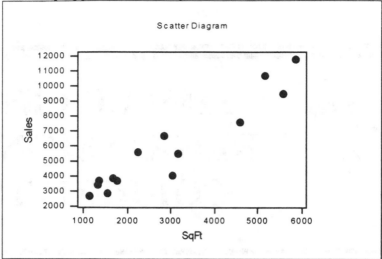

i.   If you typed rather than used the menus,

```
MTB > Plot 'Sales'*'SqFt';
SUBC>    Symbol;
SUBC>     Type 6;
SUBC>     Title "Scatter Diagram";
SUBC>      TSize 1.
```

The Least Squares Method

PROBLEM 17.8
This pertains to the previous problem 17.2. The data represent for 14 stores the square footage of the store and the annual sales. Initially, we just want to get a visual representation of the data.

a.   If the data aren't current, open the file by using the menu **File>Other Files>Import ASCII Data**; type **c1-c3** for *column* (or whatever columns you choose) and click *OK*. Give the file name *site.dat* and its location. Type the heading **Store** for c1, **SqFt** for c2, and **Sales** for c3. Most of what will be needed later for other problems will be obtained at this time.

b.   Select **Stat>Regression>Regression**

c.   For *Response*: double-click on *Sales;* for *Predictors* double-click on *SqFt*.

d.   Under *Storage* click on *Residuals*, and on *Standard.resids* and on *Fits* and on *Deleted t resids* and on *HI(leverage)* and on *Cook's Distance*. (This places some results in columns c4-c9 which will be used later.)

e.   Click on the *Options* button. Click on *Durbin-Watson statistic* and make sure the *Fit intercept* box is checked. Also, click in the box for *Prediction intervals for new observations* and type **4000** (to be used for confidence and prediction intervals). Click in the box for *Confidence level* and type **99**. Click on *OK*.

f.   Click on *OK*.

g.　The entire display is shown below.

```
Regression Analysis
The regression equation is
Sales = 901 + 1.69 SqFt

Predictor        Coef        Stdev      t-ratio         p
Constant        901.2        513.0         1.76     0.104
SqFt           1.6861       0.1533        11.00     0.000

s = 936.9        R-sq = 91.0%        R-sq(adj) = 90.2%

Analysis of Variance
SOURCE          DF           SS          MS         F          p
Regression       1     106208120   106208120    121.01     0.000
Error           12      10532255      877688
Total           13     116740376

Unusual Observations
Obs.      SqFt       Sales        Fit   Stdev.Fit    Residual    St.Resid
 14       3008        4085       5973         251       -1888       -2.09R

R denotes an obs. with a large st. resid.

Durbin-Watson statistic = 2.45

Fit    Stdev.Fit        99.0% C.I.            99.0% P.I.
7646         300     (   6729,      8562)  (    4640,     10651)
```

h.　If you typed rather than used the menus, the commands follow. Note that Minitab provides headings for the standardized residuals (*SRES1* in c4), the fitted values (*FITS1* in c5), the residuals (*RESI1* in c6), the Studentized deleted residuals (*TRES1* in c7), the hat matrix elements (*HI1* in c8), and Cook's distance statistic (*COOK1* in c9).

```
MTB > Name c4 = 'SRES1' c5 ='FITS1' c6 ='RESI1' c7 ='TRES1' c8 = 'HI1' &
CONT>        c9 = 'COOK1'
MTB > Regress 'Sales' 1 'SqFt';
SUBC>     SResiduals 'SRES1';
SUBC>     Fits 'FITS1';
SUBC>     Constant;
SUBC>     Residuals 'RESI1';
SUBC>     Tresiduals 'TRES1';
SUBC>     Hi 'HI1';
SUBC>     Cookd 'COOK1';
SUBC>     DW.
SUBC>     Predict 4000;
SUBC>        Confidence 99.
```

i.　The regression equation is shown to be Sales = 901 + 1.69 SqFt, meaning that after an overall addition of 901 thousands (or $901,000), Sales would increase by 1.69 thousands (or $1,690) for each additional square foot. The relevant portion of the display is reproduced below. (Note that the more accurate values are 901.2 and 1.6861, which are also shown in the output.)

```
Regression Analysis
The regression equation is
Sales = 901 + 1.69 SqFt
```

j.   The predicted value if the square footage is 4,000 has been provided in the output. The last two lines of the output are shown below. Note that the *Fit* or predicted sales for 4,000 square feet is 7,646 thousands, or approximately $7,646,000.

```
Fit  Stdev.Fit       99.0% C.I.          99.0% P.I.
7646        300   (    6729,    8562)  (    4640,   10651)
```

k.   There are several ways to obtain the predicted sales for a given level of square footage.

    1.   In the initial regression command we also asked for the predicted sales for 4,000 square feet, and part j. above shows the output. This is the most convenient method if you know ahead of time for what level of the independent variable you wish the predicted value.

    2.   Another approach is to use the *Predict* subcommand command directly at the session line. Part h. above shows that the *Predict* subcommand was used with the initial regression command.

    3.   A convenient approach after you have the output from the regression is to use the *Let* command as shown below.

```
MTB > let k1 = 901.2 + 1.6861 * 4000
MTB > print k1
```

        a.   The display is shown below. Sales are expected to be about $7,646 thousands ($7,646,000) when there are 4,000 square feet.

```
Data Display
K1          7645.60
```

Standard Error of the Estimate

PROBLEM 17.14
This problem continues with the Site.dat data.

    a.   The standard error is automatically provided with the least squares output. The standard error is shown below as s = 936.9 (in thousands of dollars).

```
s = 936.9        R-sq = 91.0%        R-sq(adj) = 90.2%
```

Coefficient of Determination

PROBLEM 17.20
This problem continues with the Site.dat data.

    a.   The coefficient of determination is automatically provided with the least squares output. The coefficient of determination is shown below as R-sq = 91.0 . Therefore, 91.0% of the variation in *Sales* can be explained by the variation in *square footage*. In other words, of the total variation in the *Sales* values, 91% of it is "removed" or accounted for when the least squares predictor is used rather than just using average *Sales*.

```
s = 936.9        R-sq = 91.0%        R-sq(adj) = 90.2%
```

Correlation

PROBLEM 17.29
This problem continues with the Site.dat data.
    a.   The correlation coefficient is the square root of the coefficient of variation (which is automatically provided with the least squares output) with the sign of the slope attached. There are several ways to obtain the correlation coefficient.
    b.   To obtain the correlation coefficient for the *Sales* and *SqFt* data
        1.   Select **Stat>Basic Statistics>Correlation**
        2.   Double-click on the *Sales* column provided in the list, and double-click on the *SqFt* column.
        3.   Click on *OK*. The result is shown below. The correlation coefficient is 0.954.

```
Correlations (Pearson)
Correlation of Sales and SqFt = 0.954
```

        4.   If you typed rather than used the menus,

```
MTB > Correlation 'Sales' 'SqFt'.
```

    c.   A less convenient way to obtain the correlation coefficient is shown next, only as practice in using the *let* command. Recall that the correlation coefficient is the square root of the coefficient of variation (which is 0.91) with the sign of the slope attached (and the slope is 1.69).
        1.   At the command line type

```
MTB > let k1 = sqrt(signs(1.69)*.91)
SUBC> print k1 .
```

        2.   The result is shown below. Of course, the command *signs* isn't needed in this case, since the slope 1.69 is positive. It is included just as a reminder that the correlation coefficient takes on the sign of the slope.

```
Data Display
K1        0.953939
```

Regression Diagnostics: Residual Analysis

PROBLEM 17.37
This problem continues with the Site.dat data.

Residuals versus Xs
    a.   One useful graph plots the residuals (the differences in the actual Y values and the *predicted Y* values) on the vertical axis versus the corresponding X_i values along the horizontal axis. The residuals for this problem were automatically stored by Minitab in column c6 named *RESI1* (because we checked the appropriate box when the regression command was initially executed). Therefore, since the X values are in *SqFt* and the residuals are in *RESI1*, the plot may be obtained as follows.
    b.   Select **Graph>Plot**
    c.   For *Graph Variables*: for Y double-click on *RESI1;* for X double-click on *SqFt*.
    d.   For *Display*, choose *Symbol*, and for *For each*, choose *Graph*.
    e.   Click on *Symbol* and click on the *Edit Attributes* button and under *Type* click on the arrow and select *Solid circle*, and click *OK*.

f.   Title: Click on **Annotation>Title** and type **Residuals versus Xs**
     Click in the first box under *Text Size* and type **1** to reduce the title size. Click *OK*.

g.   Click on *OK* to obtain the graph.

h.   Since the values appear random (the one value at about 3,000 square feet is a bit out of position, though), this suggests that the linear model we used is appropriate, and thus *Sales* and *SqFt* are linearly related.

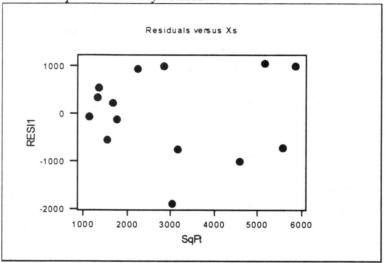

i.   If you typed rather than used the menus,

```
MTB > Plot 'RESI1'*'SqFt';
SUBC>    Symbol;
SUBC>      Type 6;
SUBC>    Title "Residuals versus Xs";
SUBC>      TSize 1;
SUBC>    Tick 1;
SUBC>    Tick 2.
```

Standardized Residuals versus Xs

This problem continues with the Site.dat data.

a.   One useful graph plots the standardized residuals (the differences in the actual *Y* values and the *predicted Y* values, divided by their standard error) on the vertical axis versus the corresponding X_i values along the horizontal axis.   The standardized residuals for this problem were automatically stored by Minitab in column c4 named *SRES1*.   Therefore, since the *X* values are in *SqFt* and the residuals are in *SRES1*, the plot may be obtained as follows.

b.   Select **Graph>Plot**

c.   For *Graph Variables*: for *Y* double-click on *SRES1;* for X double-click on *SqFt.*

d.   For *Display*, choose *Symbol*, and for *For each*, choose *Graph.*

e.   Click on *Symbol* and click on the *Edit Attributes* button and under *Type* click on the arrow and select *Solid circle*, and click *OK*.

f.   Title: Click on **Annotation>Title** and type **Standardized Residuals versus Xs**
     Click in the first box under *Text Size* and type **1** to reduce the title size. Click *OK*.
g.   Click on *OK* to obtain the graph.
h.   Since the values appear random (except for the one value at about 3,000 square feet),
     this suggests that the linear model we used is appropriate, and thus *Sales* and *SqFt* are
     linearly related.

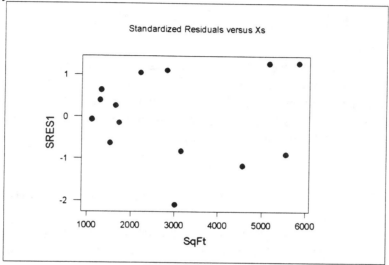

Minitab provides a suite of four plots for regression diagnostics. They are displayed here for your
information

   a.   **Stat>Regression>Residual Plots...**
   b.   For *Residuals*: double-click on *RESI1*; for *Fits* double-click on *FITS1*.
   c.   Click on *OK*.

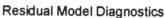

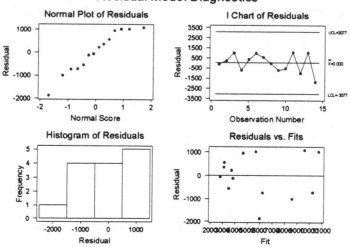

d.   If you typed rather than used the menus,
```
MTB > %Resplots 'RESI1' 'FITS1'.
```


Measuring Autocorrelation: The Durbin-Watson Statistic
This statistic helps determine whether the residuals are independent.
   a.   This value is provided with the output shown previously.
```
Durbin-Watson statistic = 2.45
```

   b.   This value is not appropriate for the *Site* data we've been analyzing, since the observations are not in any sense sequential. It is shown here only as an example of the output from Minitab. For those cases where the data *are* sequential, a value above 2 often indicates the residuals are dependent, and in such a case the regression methods we've been using may be inappropriate.

Confidence-Interval Estimate for Predicting μ_{YX} and for Predicting an Individual Response Y_i
For some specific value of X we would like to predict the mean of the Y values. In addition, we may wish to predict an individual Y value for a specific value of X. Predicting a mean is easier than predicting an individual response, since the mean has many values surrounding it. Therefore, for a specific value of X the variation in predicting an individual response is greater than for estimating the mean, and the confidence limits for predicting an individual response are outside or beyond the confidence limits for the mean (assuming the same level of confidence).

PROBLEMS 17.48 and 17.54
Both problems pertain to the *Site* data. Problem 17.48 asks for a 99% confidence interval for estimating the unknown population mean of *Sales*, while problem 17.54 asks for a 99% confidence interval for estimating an individual value. In both cases, the *SqFt* value is 4,000. The output shown below was obtained from the initial Minitab command (with these instructions: click in the box for *Prediction intervals for new observations* and type **4000**. Click in the box for *Confidence level* and type **99**).

   a.   You can see that for estimating the mean sales when *SqFt* is 4,000, the 99% confidence interval is (6729 , 8562) which is in thousands, and for estimating the individual sales value when *SqFt* is 4,000 the 99% prediction interval is (4640 , 10651) which is again in thousands.
```
Fit   Stdev.Fit      99.0% C.I.            99.0% P.I.
7646         300   (   6729,    8562)  (   4640,   10651)
```
   b.   The plot shown below provides the 99% confidence interval and the 99% prediction interval bands for *SqFt*. The specific case of *SqFt* = 4,000 is highlighted. This plot was obtained by the steps shown below.
   c.   **Stat>Regression>Fitted Line Plot...**
   d.   For *Response(Y)*: double-click on *Sales;* for *Predictor(X)* double-click on *SqFt.*

e.   Click in the box for *Confidence level* and type **99.**
f.   For *Display Options* click on *Display confidence bands* and click on *Display prediction bands*.
g.   Click on *Title* and type **Confidence and Prediction Intervals**.
h.   Click on *OK*. (Note that this graph has been edited using the Minitab graph editor, to include additional information.)

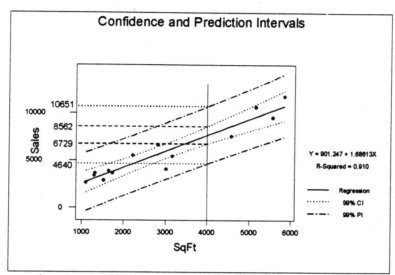

i.   If you typed rather than used the menus,

```
MTB > %Fitline 'Sales' 'SqFt';
SUBC>   Confidence 99;
SUBC>   Ci;
SUBC>   Pi;
SUBC>    Title 'Confidence and Prediction Intervals'.
```

Inferences about the population slope β_1

It would be useful to determine whether the population slope is 0. If so, then regression is not needed (since the X variable doesn't affect the Y). In addition, if the population slope is 0, then further analyses such as confidence intervals and hypothesis testing are not useful nor reliable.

PROBLEM 17.60

This problem continues with the Site.dat data.

a.   The hypotheses are H_0: $\beta_1 = 0$ versus H_1: $\beta_1 \neq 0$.

b.   This is a t-test. We have the sample t value from the initial display of the results: $t_{sample} = 11.0$, as reproduced below. For the independent variable *SqFt*, the t_{sample} is the t-ratio shown below. We need do no more work for this test, since the p-value is shown to be 0. Since this p-value is much less than any reasonable α of 0.05 or so, we are led by the sample to reject H_0 and conclude that the population slope β_1 is not zero.

```
Predictor        Coef        Stdev      t-ratio          p
Constant        901.2       513.0         1.76      0.104
SqFt           1.6861      0.1533        11.00      0.000
```

c.   If we wished, we could compare the $t_{sample} = 11.0$ with a $t_{critical}$ (which is unnecessary since we've decided to reject H_0). There are n-2 = 14-2 = 12 degrees of freedom. The level of significance is 0.01. Therefore, to find the critical t values, type

```
MTB > invcdf .995;
SUBC> t 12 df.
```

d.   The display is shown below. Therefore, the critical t values are ±3.0546. Since $t_{sample} = 11.0 > t_{critical} = 3.0546$, we are led by the sample to reject H_0, which coincides with the previous decision.

```
Inverse Cumulative Distribution Function
Student's t distribution with 12 d.f.
  P( X <= x)            x
    0.9950         3.0546
```

Influence Analysis

The goal is to test whether some observations have undue influence on the analysis, opening up the possibility of discovering some errors in the data, or perhaps repeating some analyses after removing some data.

PROBLEM 17.70

This problem continues with the Site.dat data.

a.   The *hat matrix* results were obtained in the initial Minitab command. A particular *hat matrix* value is considered a candidate for a closer look if it exceeds 4/n, which for this problem is 4/14 = 0.2857. Store 13 has a value which exceeds .2857 (see the shaded cell in the table which follows, in the *hat matrix* column), so store 13 might be removed and the analysis run again.

b. The *Studentized Deleted Residuals* were also obtained in the initial Minitab command. A *Studentized Deleted Residual* value is considered a candidate for a closer look if its absolute value exceeds $t_{.95,n-p-1}$, and n-2-1 for this case is 11. Type

```
MTB > invcdf .95 ;
SUBC> t 11 df .
```

c. The display is shown below, so the t value of interest is $t_{.95,n-3} = 1.7959$. Note in the table below (in the *Studentized Deleted Residual* column) that store 14 has a Studentized Deleted Residual whose absolute value exceeds 1.7959. Therefore, store 14 might be removed and the analysis run again.

```
Inverse Cumulative Distribution Function
Student's t distribution with 11 d.f.
  P( X <= x)              x
  0.9500              1.7959
```

d. The *Cook's Distance Statistic* values were also obtained in the initial Minitab command. A *Cook's Distance Statistic* value is considered a candidate for a closer look if it exceeds $F_{.50,p,n-p}$, which for this example is $F_{.50,2,14-2}$. Therefore, type

```
MTB > invcdf .50 ;
SUBC> F 2 12 df .
```

e. The display is shown below, so the F value of interest is $F_{.50,2,14-2} = 0.7348$. Looking at the table below (in the *Cook's Distance* column), no values exceed 0.7348, so no value has an undue influence in terms of the *Cook's Distance* value.

```
Inverse Cumulative Distribution Function
F distribution with 2 d.f. in numerator and 12 d.f. in
    denominator
P( X <= x)              x
  0.5000              0.7348
```

f.   The data are shown below. The values in this table were generated by Minitab at the beginning of working with the *site.dat* data. The shaded cells pertain to observations which might exert undue influence, and therefore might be discarded and the analysis repeated.

Store	X SqFt	Y Sales	Standardized Residuals SRES1	FITS1	Residuals RESI1	Studentized Deleted Residuals TRES1	Hat Matrix HI1	Cook's Distance COOK1
1	1726	3681	-0.14764	3811.5	-130.52	-0.14149	0.109673	0.001343
2	1642	3895	0.25546	3669.9	225.12	0.24526	0.115237	0.004250
3	2816	6653	1.11186	5649.4	1003.6	1.12400	0.071725	0.047760
4	5555	9543	-0.89751	10267.7	-724.73	-0.88968	0.257108	0.139394
5	1292	3418	0.38991	3079.7	338.27	0.37570	0.142488	0.012631
6	2208	5563	1.04758	4624.2	938.77	1.05225	0.085048	0.051005
7	1313	3660	0.62738	3115.1	544.86	0.61077	0.140668	0.032216
8	1102	2694	-0.07613	2759.4	-65.37	-0.07291	0.160028	0.000552
9	3151	5468	-0.82726	6214.3	-746.26	-0.81564	0.072841	0.026883
10	1516	2898	-0.63811	3457.4	-559.43	-0.62158	0.124292	0.028896
11	5161	10674	1.28225	9603.4	1070.61	1.32153	0.205709	0.212906
12	4567	7585	-1.17306	8601.8	-1016.83	-1.19364	0.143928	0.115677
13	5841	11760	1.28826	10750	1010.04	1.32871	0.299624	0.354994
14	3008	4085	-2.09172	5973.1	-1888.14	-2.5124	0.07163	0.168791

MULTIPLE REGRESSION MODELS

This chapter extends the simple linear regression model to the case of multiple independent variables. Inferential procedures are developed, and dummy variables and curvilinear models are discussed.

PROBLEM 18.3
There are 24 months of data, where the Distribution Cost is assumed to be dependent on *Sales* and *Number of orders*.

  a.   If the data aren't current, open the file by using the menu **File>Other Files>Import ASCII Data**; type **c1-c3** for *column* (or whatever columns you choose) and click *OK*. Give the file name *warecost.dat* and its location. Type the heading **Cost** for c1, **Sales** for c2, and **Orders** for c3.

The Scatter Diagram
This is often the initial step in analyzing data. Below is a plot for each independent variable: for *Cost* against *Sales*, and for *Cost* against *Orders*.

The *Cost* against *Sales* plot may be obtained by:
  a.   Select **Graph>Plot**
  b.   For *Graph Variables*: for *Y* double-click on *Cost;* for X double-click on *Sales.*
  c.   For *Display*, choose *Symbol*, and for *For each*, choose *Graph.*
  d.   Click on *Symbol* and click on the *Edit Attributes* button and under *Type* click on the arrow and select *Solid circle*, and click *OK*.
  e.   Title: Click on **Annotation>Title** and type **Cost versus Sales**

f. Click in the first box under *Text Size* and type **1** to reduce the title size. Click *OK*.

g. Axis tick marks: Click on **Frame>Tick** and you will see a screen with the following headings. The numbers enclosed in a box are the ones to enter. In Row 1, enter **4** for the Number of Major ticks and enter **9** for the number of Minor Ticks. Then Click OK.

	Direction	Side	Positions	Number of Major	Number of Minor
1	X			4	9

h. Click on *OK* twice to obtain the graph.

i. The relationship appears to be fairly linear.

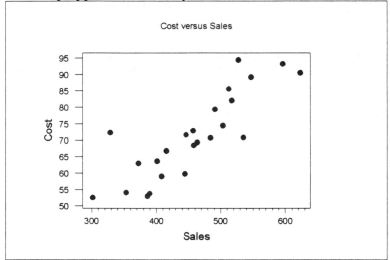

j. If you typed rather than used the menus,
```
MTB > Plot 'Cost'*'Sales';
SUBC>    Symbol;
SUBC>      Type 6;
SUBC>    Title "Cost versus Sales";
SUBC>      TSize 1;
SUBC>    Same 2;
SUBC>    Tick 1;
SUBC>      Number 4 9;
SUBC>    Tick 2.
```

The *Cost* against *Orders* plot may be obtained by:

a. Select **Graph>Plot**

b. For *Graph Variables*: for *Y* double-click on *Cost;* for X double-click on *Orders.*

c. For *Display*, choose *Symbol*, and for *For each*, choose *Graph.*

d. Click on *Symbol* and click on the *Edit Attributes* button and under *Type* click on the arrow and select *Solid circle*, and click *OK*.

e. Title: Click on **Annotation>Title** and type **Cost versus Orders**

f. Click in the first box under *Text Size* and type **1** to reduce the title size. Click *OK*.

g. Axis tick marks: Click on **Frame>Tick** and you will see a screen with the following headings. The numbers enclosed in a box are the ones to enter. In Row 1, enter **4** for the Number of Major ticks and enter **9** for the number of Minor Ticks. Then Click OK.

	Direction	Side	Positions	Number of Major	Number of Minor
1	X			4	9

h.   Click on *OK* twice to obtain the graph.

i.   The relationship appears to be fairly linear.

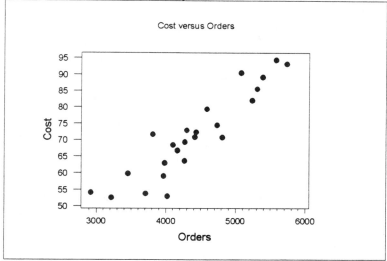

j.   If you typed rather than used the menus,

```
MTB > Plot 'Cost'*'Orders';
SUBC>    Symbol;
SUBC>      Type 6;
SUBC>    Title "Cost versus Orders";
SUBC>      TSize 1;
SUBC>    Same 2;
SUBC>    Tick 1;
SUBC>      Number 4 9;
SUBC>    Tick 2.
```

3D Plot

The following is a 3D plot of *Cost* against *Sales* and *Orders*:

a.   Select **Graph>3D Plot...**

b.   For *Graph Variables*: for *Z* double-click on *Cost;* for *Y* double-click on *Orders*, and for *X* double-click on *Sales*.

c.   For *Display*, choose *Symbol*, and for *For each*, choose *Graph*.

d.   Title: Click on **Annotation>Title** and type **Cost versus Sales and Orders**

e.   Click in the first box under *Text Size* and type **1** to reduce the title size. Click *OK*.

f.   Click on *OK* to obtain the graph.

g.   The 3D Plot is shown below.

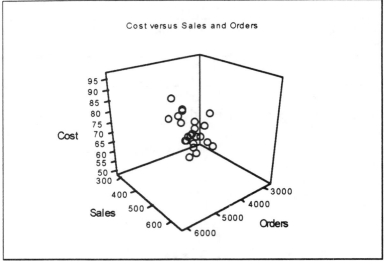

h.   If you typed rather than using the menus,

```
MTB > Pltx 'Cost'*'Orders'*'Sales';
SUBC>    Symbol;
SUBC>    Title "Cost versus Sales and Orders";
SUBC>     TSize 1;
SUBC>    HSRemoval 1.
```

Developing the Multiple Regression Model

PROBLEM 18.3, continued
There are 24 months of data, where the Distribution Cost is assumed to be dependent on Sales and Number of orders. Most of what will be needed later for other problems will be obtained at this time.

a.   Select **Stat>Regression>Regression...**
b.   For *Response*: double-click on *Cost;* for *Predictors* double-click on *Sales* and double-click on *Orders*.
c.   Under *Storage* click on *Residuals*, and on *Standard.resids* and on *Fits* and on *Deleted t resids* and on *HI(leverage)* and on *Cook's Distance*. (This places some results in columns c4-c9 which will be used later.)
d.   Click on the *Options* button. Click on *Durbin-Watson statistic* and click on *Variance inflation factors* and make sure the *Fit intercept* box is checked. Also, click in the box for *Prediction intervals for new observations* and type **400 4500** (to be used for confidence and prediction intervals: note that *Sales* are in $1,000s). Click in the box for *Confidence level* and type **95**. Click on *OK*.
e.   Click on *OK*.

f.   The entire display is shown below.

```
Regression Analysis
The regression equation is
Cost = - 2.73 + 0.0471 Sales + 0.0119 Orders

Predictor        Coef        Stdev     t-ratio          p         VIF
Constant       -2.728        6.158       -0.44      0.662
Sales         0.04711      0.02033        2.32      0.031         2.8
Orders       0.011947     0.002249        5.31      0.000         2.8

s = 4.766        R-sq = 87.6%      R-sq(adj) = 86.4%

Analysis of Variance
SOURCE         DF          SS          MS         F          p
Regression      2      3368.1      1684.0     74.13      0.000
Error          21       477.0        22.7
Total          23      3845.1

SOURCE         DF       SEQ SS
Sales           1      2726.8
Orders          1       641.3

Unusual Observations
Obs.       Sales          Cost         Fit    Stdev.Fit      Residual
     St.Resid
  1          386        52.950      63.425        1.332       -10.475        -
     2.29R

R denotes an obs. with a large st. resid.

Durbin-Watson statistic = 2.26

Fit   Stdev.Fit      95.0% C.I.            95.0% P.I.
 69.878      1.663   ( 66.419,  73.338)   ( 59.378,  80.379)
```

g.   If you typed rather than used the menus, the commands follow. Note that Minitab provides headings for the standardized residuals (*SRES1* in c4), the fitted values (*FITS1* in c5), the residuals (*RESI1* in c6), the Studentized deleted residuals (*TRES1* in c7), the hat matrix elements (*HI1* in c8), and Cook's distance statistic (*COOK1* in c9).

```
MTB > Name c4 = 'SRES1' c5 = 'FITS1' c6 = 'RESI1' c7 = 'TRES1' &
CONT>      c8 = 'HI1' c9 = 'COOK1'
MTB > Regress 'Cost' 2 'Sales' 'Orders';
SUBC>    SResiduals 'SRES1';
SUBC>    Fits 'FITS1';
SUBC>    Constant;
SUBC>    Residuals 'RESI1';
SUBC>    Tresiduals 'TRES1';
SUBC>    Hi 'HI1';
SUBC>    Cookd 'COOK1';
SUBC>    VIF;
SUBC>    DW;
SUBC>    Predict 400 4500.
```

h.   The regression equation is shown to be
        Cost = - 2.73 + 0.0471 Sales + 0.0119 Orders,
meaning that after an overall deduction of 2.73 thousands ($2,730), *Costs* would increase by 0.0471 thousands (or $47.10, since *Costs* are in $1,000s) for each 1 unit increase in the *Sales* value (with *Orders* held constant), and here *Sales* are in $1,000s. Further, *Costs* would increase by 0.0119 thousands (or $11.90, since *Costs* are in $1,000s) for each 1 unit increase in the *Orders* value (with *Sales* held constant). The relevant portion of the display is reproduced below.

```
The regression equation is
Cost = - 2.73 + 0.0471 Sales + 0.0119 Orders
```

Predicting Y

PROBLEM 18.7

i.   In order to compute the predicted value if *Sales* are 400 (thousands) and *Orders* are 4,500, type

```
MTB > let k1 = - 2.73 + 0.0471*400 + 0.0119*4500
MTB > print k1
```

j.   The display is shown below. Costs are expected to be about 69.66 thousands of dollars, or $69,660 when *Sales* are 400 (thousands) and *Orders* are 4,500.

```
Data Display
K1          69.6600
```

Coefficient of Multiple Determination

PROBLEM 18.12

k.   Continuing with the warecost.dat result, the Coefficient of Multiple Determination is automatically provided with the least squares output. The standard error $r^2_{Y.12} = .876$ is shown below as R-sq = 87.6% .

```
s = 4.766          R-sq = 87.6%          R-sq(adj) = 86.4%
```

l.   The adjusted r^2 is also automatically provided. The $r^2_{adj} = .864$, which is shown above as R-sq(adj) = 86.4% .

Residual Analysis

PROBLEM 18.16

Plotting standardized residuals versus the fits:
     A plot may be obtained by:

a.   Select **Graph>Plot**
b.   For *Graph Variables*: for *Y* double-click on *SRES1*; for X double-click on *FITS1*.
c.   For *Display*, choose *Symbol*, and for *For each*, choose *Graph*.
d.   Click on *Symbol* and click on the *Edit Attributes* button and under *Type* click on the arrow and select *Solid circle*, and click *OK*.
e.   Title: Click on **Annotation>Title** and type **Standardized Residuals versus Fits**
f.   Click in the first box under *Text Size* and type **1** to reduce the title size. Click *OK*.

g. Click on *OK* to obtain the graph.
h. Since the values appear random, this suggests that the linear model we used is appropriate, and thus *Costs* and *Sales* seem to be linearly related, and *Costs* and *Orders* seem to be linearly related.

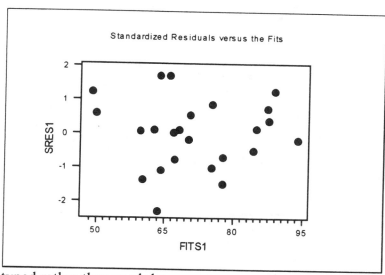

i. If you typed rather than used the menus,

```
MTB > Plot 'SRES1'*'FITS1';
SUBC>    Symbol;
SUBC>     Type 6;
SUBC>    Title "Standardized Residuals versus the Fits";
SUBC>     TSize 1;
SUBC>    Same 2;
SUBC>    Tick 1;
SUBC>     Number 4 9;
SUBC>    Tick 2.
```

Plotting standardized residuals versus the *Sales*

A plot may be obtained by:

a. Select **Graph>Plot**
b. For *Graph Variables*: for *Y* double-click on *SRES1*; for X double-click on *Sales*.
c. For *Display*, choose *Symbol*, and for *For each*, choose *Graph*.
d. Click on *Symbol* and click on the *Edit Attributes* button and under *Type* click on the arrow and select *Solid circle*, and click *OK*.
e. Title: Click on **Annotation>Title** and type **Standardized Residuals versus Sales**
f. Click in the first box under *Text Size* and type **1** to reduce the title size. Click *OK*.
g. Click on *OK* to obtain the graph.

h.   Since the values appear random, this suggests that the linear model we used is appropriate, and thus *Costs* and *Sales* are linearly related.

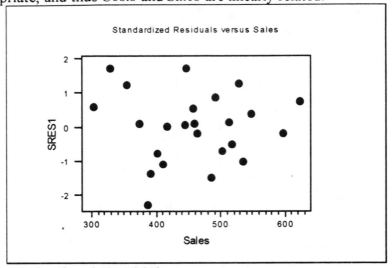

i.   If you typed rather than used the menus,
```
MTB > Plot 'SRES1'*'Sales';
SUBC>    Symbol;
SUBC>      Type 6;
SUBC>    Title "Standardized Residuals versus Sales";
SUBC>      TSize 1;
SUBC>    Same 2;
SUBC>    Tick 1;
SUBC>      Number 4 9;
SUBC>    Tick 2.
```

Plotting standardized residuals versus the *Costs*
a.   Select **Graph>Plot**
b.   For *Graph Variables*: for Y double-click on *SRES1*; for X double-click on *Orders*.
c.   For *Display*, choose *Symbol*, and for *For each*, choose *Graph*.
d.   Click on *Symbol* and click on the *Edit Attributes* button and under *Type* click on the arrow and select *Solid circle*, and click *OK*.
e.   Title: Click on **Annotation>Title** and type **Standardized Residuals versus Orders**
f.   Click in the first box under *Text Size* and type **1** to reduce the title size. Click *OK*.
g.   Click on *OK* to obtain the graph.

h.   Since the values appear random, this suggests that the linear model we used is appropriate, and thus *Costs* and *Orders* are linearly related.

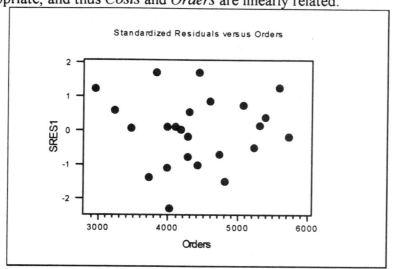

i.   If you typed rather than used the menus,

```
MTB > Plot 'SRES1'*'Orders';
SUBC>    Symbol;
SUBC>      Type 6;
SUBC>    Title "Standardized Residuals versus Orders";
SUBC>      TSize 1;
SUBC>    Same 2;
SUBC>    Tick 1;
SUBC>      Number 4 9;
SUBC>    Tick 2.
```

Plotting standardized residuals versus the months.
For this plot we'll use a different Minitab menu item.

a.   Select **Graph>Time Series Plot...**

b.   For *Graph Variables*: for *Y* double-click on *SRES1*

c.   For *Time Scale* click on *Calendar* and use the arrow to select *Months*

d.   Click on *Symbol* and click on the *Edit Attributes* button and under *Type* click on the arrow and select *Solid circle*, and click *OK*.

e.   Title: Click on **Annotation>Title** and type **Standardized Residuals versus Months**

f.   Click in the first box under *Text Size* and type **1** to reduce the title size. Click *OK*.

g.   Click on *OK* to obtain the graph.

h.   The purpose of this test is to determine whether the residuals are independent. There should be no pattern in the values below, and it seems that there is no clear pattern. (Of interest, however, is that the first year's pattern is somewhat duplicated in the second year.)

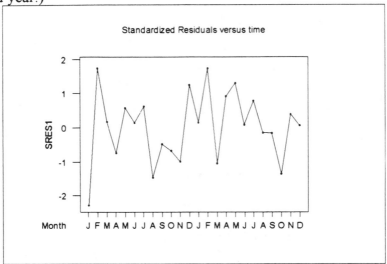

i.   If you typed rather than used the menus,

```
MTB > TSPlot 'SRES1';
SUBC>    Month;
SUBC>    TDisplay 11;
SUBC>    Symbol;
SUBC>      Type 6;
SUBC>    Connect;
SUBC>    Title "Standardized Residuals versus time";
SUBC>      TSize 1;
SUBC>    Tick 11;
SUBC>    Tick 2.
```

Measuring Autocorrelation: The Durbin-Watson Statistic

This statistic helps determine whether the residuals are independent.

a.   This value is provided with the output shown previously.
```
Durbin-Watson statistic = 2.26
```

b.   This value is appropriate for the *Warecost* data we've been analyzing, since the observations are sequential.

c.   To test whether autocorrelation is present: compare the $DW_{sample} = 2.26$ with the critical values *upper* $DW_{critical}$ and *lower* $DW_{critical}$. The rule is stated in several parts: (1) if $DW_{sample} <$ *lower* $DW_{critical}$ then reject the null hypothesis and conclude that there *is* evidence of positive autocorrelation; (2) If $DW_{sample} >$ *upper* $DW_{critical}$ then we may not reject the null hypothesis, and therefore we conclude there is *no* evidence of positive autocorrelation; (3) If *lower* $DW_{critical} \leq DW_{sample} \leq$ *upper* $DW_{critical}$ then the test is inconclusive.

    d.   From Table E14, for n = 24, the *upper* DW$_{critical}$ is 1.55 and the *lower* DW$_{critical}$ is 1.19. Since DW$_{sample}$ = 2.26 > *upper* DW$_{critical}$ = 1.55, do not reject the null hypothesis. Therefore, we conclude that the autocorrelation parameter is zero, and that there is no evidence of positive autocorrelation. We assume, then, that the error terms are either uncorrelated random variables, or that they are independent normal random variables.

Inferences about the population slope β_1 and β_2

It would be useful to determine whether for each parameter the population slope is 0. If so, then regression is not needed, and we cannot rely on the results of regression analysis.

PROBLEM 18.21

Continuing with the *Warecost* data. Test at α = .05 whether the *Sales* slope is zero, and determine the p-value.

    a. H_0: $\beta_1 = 0$ versus H_1: $\beta_1 \neq 0$.

    b.   This is a t-test. We have the sample t value from the initial display of the results: t_{sample} = 2.32, which is partially reproduced below. For the independent variable *Sales*, the t_{sample} is the t-ratio shown below. We need do no more work for this test, since the p-value for *Sales* is shown to be 0.031. Since this p-value is much less than any reasonable α of 0.05 or so, we are led by the sample to reject H_0 and conclude that the population slope β_1 (*Sales*) is not zero.

```
Predictor       Coef      Stdev     t-ratio        p        VIF
Constant       -2.728      6.158      -0.44      0.662    ·
Sales          0.04711    0.02033     2.32       0.031       2.8
Orders         0.011947   0.002249    5.31       0.000       2.8
```

    c.   If we wished, we could compare the t_{sample} = 2.32 with a $t_{critical}$ (which is unnecessary since we've decided to reject H_0). There are n-2-1 = 24-2-1 = 21 degrees of freedom. The level of significance is 0.05. Therefore, to find the critical t values, type

```
MTB > invcdf .975;
SUBC> t 21 df.
```

    d.   The display is shown below. Therefore, the critical t values are ± 2.0796. Since t_{sample} = 2.32 > $t_{critical}$ = 2.0796, we are led by the sample to reject H_0, which coincides with the previous decision.

```
Inverse Cumulative Distribution Function
Student's t distribution with 21 d.f.
  P( X <= x)            x
     0.9750         2.0796
```

Continuing with the *Warecost* data. Test at α = .05 whether the *Orders* slope is zero, and determine the p-value.

    e. H_0: $\beta_2 = 0$ versus H_1: $\beta_2 \neq 0$.

    f.   This is a t-test. We have the sample t value from the initial display of the results: t_{sample} = 5.31, as shown below. For the independent variable *Orders*, the t_{sample} is the t-ratio shown below. We need do no more work for this test, since the p-value for

Orders is shown to be 0.000. Since this p-value is much less than any reasonable α of 0.05 or so, we are led by the sample to reject H_0 and conclude that the population slope β_2 *(Orders)* is not zero.

Predictor	Coef	Stdev	t-ratio	p	VIF
Constant	-2.728	6.158	-0.44	0.662	
Sales	0.04711	0.02033	2.32	0.031	2.8
Orders	0.011947	0.002249	5.31	0.000	2.8

g. If we wished, we could compare the $t_{sample} = 5.31$ with a $t_{critical}$ (which is unnecessary since we've decided to reject H_0). There are n-2-1 = 24-2-1 = 21 degrees of freedom. The level of significance is 0.05. Therefore, to find the critical t values, type

```
MTB > invcdf .975;
SUBC> t 21 df.
```

h. The display is shown below. Therefore, the critical t values are ±2.0796. Since $t_{sample} = 5.31 > t_{critical} = 2.0796$, we are led by the sample to reject H_0, which coincides with the previous decision.

```
Inverse Cumulative Distribution Function
Student's t distribution with 21 d.f.
  P( X <= x)             x
    0.9750            2.0796
```

Testing Portions of a Multiple Regression Model

PROBLEM 18.26

The goal is to test the contribution of each independent variable, using the partial F test criterion. Minitab provides the information to perform these tests. The Regression sum of squares when both Sales and Orders were included in the model was 3368.1 (that is, the amount of total variation removed because both were in the model), and this value is shown in the F table reproduced below, which was the output from the initial commands for this chapter. Below the F table is a mini-table which has the heading **SEQ SS** for its third column.

a. This is the F table. This is part of the original output.

SOURCE	DF	SS	MS	F	p
Regression	2	3368.1	1684.0	74.13	0.000
Error	21	477.0	22.7		
Total	23	3845.1			

b. This is the Sequential sums of squares table when *Orders given Sales* is in the model. This is part of the original output.

SOURCE	DF	SEQ SS
Sales	1	2726.8
Orders	1	641.3

c.　This is the Sequential sums of squares table when *Sales given Orders* is in the model. This is *not* part of the original output. This information was obtained when the initial least squares Minitab commands were used, but with the order reversed; that is, this model was for Costs given that Orders was the first independent variable, and Sales was the second.

```
SOURCE          DF       SEQ SS
Orders          1        3246.1
Sales           1         122.0
```

That is, the following was used, which has *Orders* and *Sales* in a different order.
1.　Select **Stat>Regression>Regression**
2.　For *Response*: double-click on *Cost;* for *Predictors* double-click on *Orders* and then double-click on *Sales*.

There are two tests to perform, using the SEQ SS information.

First: Testing *Sales* given *Orders is in the model*
a.　Test at $\alpha=.05$: H_0: Sales *does not* significantly improve the model once *Orders* is already in, versus H_1: Sales *does* improve it.
b.　To compute the F value, look at the second *SEQ SS* table, where the SEQ SS for Sales with Orders already in is 122.0. This has 1 df, and the error mean squares is 22.7 . Type
```
MTB > let k1 = 122/22.7
MTB > print k1
```
c.　The display is shown
```
Data Display
K1       5.37445
```
d　Is this F_{sample} = 5.37445 significant? Compute the $F_{critical}$ by typing
```
MTB > invcdf .95;
SUBC> f 1 21 .
```
e.　The display is shown. Since F_{sample} = 5.37445 > $F_{critical}$ = 4.3249, reject H_0. Therefore, Sales significantly improves the model once Orders is already in.
```
Inverse Cumulative Distribution Function
F distribution with 1 d.f. in numerator and 21 d.f. in
    denominator
  P( X <= x)          x
    0.9500         4.3249
```
Second: Testing *Orders* given *Sales is in the model*
a.　Test at $\alpha=.05$: H_0: Orders *does not* significantly improve the model once *Sales* is already in, versus H_1: Orders *does* improve it.
b.　To compute the F value, look at the first *SEQ SS* table, where the SEQ SS for Orders with Sales already in is 641.3. This has 1 df, and the error mean squares is 22.7 . Type
```
MTB > let k1 = 641.3/22.7
MTB > print k1
```

c. The display is shown
```
Data Display
K1        28.2511
```

d. Is this $F_{sample} = 28.2511$ significant? The $F_{critical}$ was already determined to be 4.3249. Since $F_{sample} = 28.2511 > F_{critical} = 4.3249$, reject H_0. Therefore, Orders significantly improves the model once Sales is already in.

Confidence-Interval Estimation of the slope

PROBLEM 18.31
Continuing with the *warecost* data. Construct a 95% confidence interval for estimating the unknown population slope between *Costs* and *Sales*. This may be computed from the Minitab output displayed at the beginning of the chapter. The form of the confidence interval (since *Sales* pertains to X_1) is $b_1 \pm t_{n-p-1} S_{b_1}$.

a. From the output reproduced below, we have $b_1 \pm t_{n-p-1} S_{b_1} = 0.04711 \pm t_{21}(0.02033)$.

Predictor	Coef	Stdev	t-ratio	p	VIF
Constant	-2.728	6.158	-0.44	0.662	
Sales	0.04711	0.02033	2.32	0.031	2.8
Orders	0.011947	0.002249	5.31	0.000	2.8

b. To find the correct t value, type
```
MTB > invcdf .975 ;
SUBC> t 21 .
```

c. The display is shown below.
```
Inverse Cumulative Distribution Function
Student's t distribution with 21 d.f.
  P( X <= x)            x
    0.9750           2.0796
```

d. If you want Minitab to compute the upper and lower limits, type
```
MTB > let k1 = .04711 + 2.0796*.02033
MTB > let k2 = .04711 - 2.0796*.02033
MTB > print k1 k2
```

e. The display is shown below. Therefore, $b_1 \pm t_{n-p-1} S_{b_1} = 0.04711 \pm t_{21}(0.02033) = 0.04711 \pm 2.0796(0.02033) = (0.0048, 0.0894)$.
```
Data Display
K1        0.0893883
K2        0.00483173
```

Coefficient of Partial Determination
PROBLEM 18.36
This continues with the problem from 18.3, the *warecost* data.

   a.   Computation of these coefficients uses the results from an earlier section, *Testing Portions of a Multiple Regression Model* (PROBLEM 18.26). The two coefficients are

$$r^2_{Y1.2} = \frac{SSR(X_1 \mid X_2)}{SST - SSR(X_1 \cap X_2) + SSR(X_1 \mid X_2)}$$

$$\text{and} \quad r^2_{Y2.1} = \frac{SSR(X_2 \mid X_1)}{SST - SSR(X_1 \cap X_2) + SSR(X_2 \mid X_1)} .$$

   b.   Some of the display from the problem is shown below. This is the F table. This is part of the original output.

```
SOURCE        DF          SS          MS        F        p
Regression     2       3368.1      1684.0     74.13    0.000
Error         21        477.0        22.7
Total         23       3845.1
```

   c.   This is the Sequential sums of squares table for Orders given Sales is in the model. This is part of the original output.

```
SOURCE        DF        SEQ SS
Sales          1        2726.8
Orders         1         641.3
```

   d.   This is the Sequential sums of squares table for Sales given Orders is in the model. This is *not* part of the original output. This information was obtained when the initial least squares Minitab commands were used, but with the order reversed; that is, this model was for Costs given that Orders was the first independent variable, and Sales was the second.

```
SOURCE        DF        SEQ SS
Orders         1        3246.1
Sales          1         122.0
```

   e.   Some of the output shown above is needed for this problem:

   f.   From the F table of the original display: SST = 3845.1, and $SSR(X_1 \cap X_2) = 3368.1$.

   g.   From the results of the section *Testing Portions of a Multiple Regression Model*: $SSR(X_1 \mid X_2) = 122$, and $SSR(X_2 \mid X_1) = 641.3$. Therefore, to compute the two values, type

```
MTB > let k1 = 122/(3845.1-3368.1+122)
MTB > let k2 = 641.3/(3845.1-3368.1+641.3)
MTB > print k1 k2
```

   h.   The display is shown below. Therefore, $r^2_{Y1.2} = 0.203673$ and $r^2_{Y2.1} = 0.573460$.

```
Data Display
K1        0.203673
K2        0.573460
```

The Curvilinear Regression Model

PROBLEM 18.41
A psychologist wishes to predict the number of typing errors based upon the amount of alcoholic consumption.

a.   Open the file by using the menu **File>Other Files>Import ASCII Data**; type **c1-c2** for *column* (or whatever columns you choose) and click *OK*. Give the file name *alcohol.dat* and its location. Type the heading **Oz** for c1 and **Errors** for c2.

b.   Since this is a curvilinear model, we need another term; an X^2 term. Therefore, type in c3 the heading **OzSq** for squared errors. To place the squares of the *Oz* values into the *OzSq* column, type
MTB > `let 'OzSq' = 'Oz'**2`

c.   Select **Stat>Regression>Regression...**

d.   For *Response*: double-click on *Errors*; for *Predictors* double-click on *Oz* and double-click on *OzSq*.

e.   Under *Storage* click on *Residuals*, and on *Standard.resids* and on *Fits* and on *Deleted t resids* and on *HI(leverage)* and on *Cook's Distance*. (This places some results in columns c4-c9 which will be used later.)

f.   Click on the *Options* button. Click on *Durbin-Watson statistic* and click on *Variance inflation factors* and make sure the *Fit intercept* box is checked. Also, click in the box for *Prediction intervals for new observations* and type **2.5 6.25** (to be used for confidence and prediction). Click in the box for *Confidence level* and type **95**. Click on *OK*.

g.   Click on *OK*.

h.   The entire display is shown below.

```
Regression Analysis

The regression equation is
Errors = 4.18 + 0.44 Oz + 1.19 OzSq

Predictor        Coef        Stdev      t-ratio          p          VIF
Constant        4.181        1.435         2.91      0.013
Oz              0.438        1.700         0.26      0.801         12.4
OzSq           1.1905       0.4075         2.92      0.013         12.4

s = 2.641        R-sq = 91.2%        R-sq(adj) = 89.8%

Analysis of Variance
SOURCE        DF            SS           MS          F          p
Regression     2        870.72       435.36      62.44      0.000
Error         12         83.68         6.97
Total         14        954.40

SOURCE        DF        SEQ SS
Oz             1        811.20
OzSq           1         59.52
```

```
Unusual Observations
Obs.       Oz      Errors        Fit   Stdev.Fit   Residual     St.Resid
 14       4.00     30.000     24.981      1.435       5.019        2.26R

R denotes an obs. with a large st. resid.

Durbin-Watson statistic = 2.58

      Fit   Stdev.Fit        95.0% C.I.           95.0% P.I.
   12.717      1.016     ( 10.503,  14.930)   (  6.551,  18.883)
```

i. If you typed rather than used the menus, the commands follow. Note that Minitab provides headings for the standardized residuals (*SRES1* in c4), the fitted values (*FITS1* in c5), the residuals (*RESI1* in c6), the Studentized deleted residuals (*TRES1* in c7), the hat matrix elements (*HI1* in c8), and Cook's distance statistic (*COOK1* in c9). Notice that Minitab typed the headings for some new columns which will be used later.

```
MTB > Name c4 = 'SRES1' c5 = 'FITS1' c6 = 'RESI1' c7 = 'TRES1' c8 =
      'HI1' &
CONT>      c9 = 'COOK1'
MTB > Regress 'Errors' 2  'Oz' 'OzSq';
SUBC>    SResiduals 'SRES1';
SUBC>    Fits 'FITS1';
SUBC>    Constant;
SUBC>    Residuals 'RESI1';
SUBC>    Tresiduals 'TRES1';
SUBC>    Hi 'HI1';
SUBC>    Cookd 'COOK1';
SUBC>    VIF;
SUBC>    DW;
SUBC>    Predict 2.5 6.25.
```

The Scatter Diagram
This is often the initial step in analyzing data. Below is a plot for *Errors* against *Oz*.

Plot for *Errors* against *Oz*.
a. Select **Graph>Plot...**
b. For *Graph Variables*: for Y double-click on *Errors*; for X double-click on *Oz*.
c. For *Display*, choose *Symbol*, and for *For each*, choose *Graph*.
d. Click on *Symbol* and click on the *Edit Attributes* button and under *Type* click on the arrow and select *Solid circle*, and click *OK*.
e. Title: Click on **Annotation>Title** and type **Errors versus Oz**
f. Click in the first box under *Text Size* and type **1** to reduce the title size. Click *OK*.
g. Click on *OK* to obtain the graph.

h.   The relationship appears to be curvilinear, which is supported by the test for a curvilinear relationship in part o shown below.

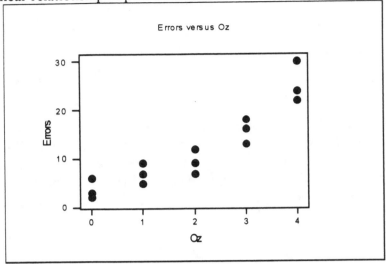

i.   If you typed rather than used the menus,
```
MTB > Plot 'Errors'*'Oz';
SUBC>    Symbol;
SUBC>     Type 6;
SUBC>    Title "Errors versus Oz";
SUBC>     TSize 1;
SUBC>    Tick 1;
SUBC>    Tick 2.
```

j.   The regression equation is shown to be *Errors = 4.18 + 0.44 Oz + 1.19 OzSq*, meaning that after an overall value of 4.18, the number of *Errors* would increase by 0.044 for each 1 *Oz* value (with *OzSq* held constant). Further, the number of *Errors* would increase 1.19 for each 1 *OzSq* value (with *Oz* held constant). The relevant portion of the display is reproduced below.
```
The regression equation is
Errors = 4.18 + 0.44 Oz + 1.19 OzSq
```

k.   To predict the number of *Errors* if the person consumes 2.5 *Oz* (which means 2.5*2.5 = 6.25 *OzSq* as well), use the regression equation shown above. Type
```
MTB > let k1 = 4.18 + .44*2.5 + 1.19*6.25
MTB > print k1
```

l.   The display is shown. The predicted number of *Errors* when 2.5 *Oz* are consumed is 12.7175.
```
Data Display
K1       12.7175
```

m. To determine whether there is a significant curvilinear relationship between alcoholic consumption and the number of errors, at the $\alpha=.05$ level of significance, we use the F table which is reproduced below.

```
SOURCE        DF          SS         MS         F        p
Regression    2        870.72     435.36     62.44    0.000
Error        12         83.68       6.97
Total        14        954.40
```

n. The F_{sample} is shown to be 62.44. The $F_{critical}$ for p=2 and n-p=12 degrees of freedom may be determined from Minitab by typing

```
MTB > invcdf .95;
SUBC> f 2 12 df .
```

o. The display is shown. Therefore, since $F_{sample} = 62.44 > F_{critical} = 3.8853$, reject H_0 that there is no significant relationship. Therefore, we are led by the sample to conclude that there is a significant curvilinear relationship between alcoholic consumption and the number of errors.

```
Inverse Cumulative Distribution Function
F distribution with 2 d.f. in numerator and 12 d.f. in denominator
P( X <= x)           x
   0.9500         3.8853
```

p. The Coefficient of Multiple Determination is automatically provided with the least squares output. The standard error $r^2_{Y.12} = 0.912$ is reproduced below as R-sq = 91.2%. Therefore 91.2% of the variation in *Errors* can be explained by the curvilinear relationship with consumption, *Oz*.

```
s = 2.641        R-sq = 91.2%        R-sq(adj) = 89.8%
```

q. The adjusted r^2 is also automatically provided. $r^2_{adj} = 0.898$, which is shown above as R-sq(adj) = 89.8%.

r. Is the curvilinear model effective, or would a simple model without the squared *OzSq* term be just as effective? Test this at $\alpha=.05$.

s. H_0: The curvilinear model is no better than the simple model (the slope for *OzSq* = 0) versus

H_1: The curvilinear model is an improvement.

t. The pertinent information from the output is reproduced below. The slope of *OzSq* is 1.1905, and its t_{sample} (shown as t-ratio below) = 2.92. We need do no more work, since the p-value is 0.013, which is much smaller than any reasonable value for α, such as 0.05. Therefore we would reject H_0 and conclude that the curvilinear model is effective.

```
Predictor       Coef        Stdev       t-ratio        p         VIF
Constant        4.181       1.435        2.91       0.013
Oz              0.438       1.700        0.26       0.801       12.4
OzSq            1.1905      0.4075       2.92       0.013       12.4
```

u. If we wished to find the $t_{critical}$ value, note that there are n-2 degrees of freedom, so type

```
MTB > invcdf .975;
SUBC> t 12 df .
```

v.   The display is shown. Since $t_{sample} = 2.92 > t_{critical} = 2.1788$, then we are again led by the sample to reject H_0. The curvilinear model is effective.

```
Inverse Cumulative Distribution Function
Student's t distribution with 12 d.f.
P( X <= x)              x
    0.9750           2.1788
```

Dummy-Variable Models

In Minitab it is very easy to include a dummy variable, which is then treated like any other independent variable.

PROBLEM 18.46

This problem pertains to the amount of money withdrawn from ATM machines as it relates to the median prices of the nearby neighborhood. There is a dummy column containing a 1 if the ATM is located at a shopping center, and a 0 if it is not.

a.   Open the file by using the menu **File>Other Files>Import ASCII Data**; type **c1-c3** for *column* (or whatever columns you choose) and click *OK*. Give the file name *atm2.dat* and its location. Type the heading **Cash** for c1, **Home** for c2, and **Locat** for c3.

b.   Select **Stat>Regression>Regression...**

c.   For *Response*: double-click on *Cash;* for *Predictors* double-click on *Home* and double-click on *Locat*.

d.   Under *Storage* click on *Residuals*, and on *Standard.resids* and on *Fits* and on *Deleted t resids* and on *HI(leverage)* and on *Cook's Distance*. (This places some results in columns c4-c9 which could be used later.)

e.   Click on the *Options* button. Click on *Durbin-Watson statistic* and make sure the *Fit intercept* box is checked. Click on *OK*.

f.   Click on *OK*.

g.   The entire display is shown below.

```
Regression Analysis

The regression equation is
Cash = 29.7 + 0.393 Home + 1.23 Locat

Predictor        Coef       Stdev    t-ratio        p       VIF
Constant       29.682       1.334      22.25    0.000
Home         0.393129    0.007346      53.52    0.000       1.1
Locat          1.2282      0.5459       2.25    0.044       1.1

s = 0.9990      R-sq = 99.6%      R-sq(adj) = 99.6%
```

```
Analysis of Variance
SOURCE        DF          SS          MS          F          p
Regression     2       3278.4      1639.2     1642.59     0.000
Error         12         12.0         1.0
Total         14       3290.4

SOURCE        DF      SEQ SS
Home           1      3273.4
Locat          1         5.1

Unusual Observations
Obs.     Home       Cash        Fit   Stdev.Fit   Residual   St.Resid
  2       170     99.000     96.514      0.379      2.486       2.69R

R denotes an obs. with a large st. resid.

Durbin-Watson statistic = 1.93
```

h.   If you typed rather than used the menus, the commands follow. Note that Minitab provides headings for the standardized residuals (*SRES1* in c4), the fitted values (*FITS1* in c5), the residuals (*RESI1* in c6), the Studentized deleted residuals (*TRES1* in c7), the hat matrix elements (*HI1* in c8), and Cook's distance statistic (*COOK1* in c9). Notice that Minitab typed the headings for some new columns.

```
MTB > Name c4 = 'SRES1' c5 = 'FITS1' c6 = 'RESI1' c7 = 'TRES1' c8 =
      'HI1' &
CONT>      c9 = 'COOK1'
MTB > Regress 'Cash' 2 'Home' 'Locat';
SUBC>    SResiduals 'SRES1';
SUBC>    Fits 'FITS1';
SUBC>    Constant;
SUBC>    Residuals 'RESI1';
SUBC>    Tresiduals 'TRES1';
SUBC>    Hi 'HI1';
SUBC>    Cookd 'COOK1';
SUBC>    VIF.
SUBC>    DW.
```

i.   The output would be used in the same way we've been using it; to answer various questions. The dummy column is treated by Minitab as just another independent variable.

Regression Models Using Transformations

Transformations are very easy to accomplish in Minitab. As we saw in the case where we added a new squared term by typing `MTB > let 'OzSq' = 'Oz'**2` , which added the new column with squared values. In this manner new columns may be created which have values which are square roots of other values, or logs, and so on. Once the columns have been generated, then Minitab will treat them like any other columns in its regression analyses. Including an interaction term is an example of this process, where an interaction term consists of a new column made up of products of the corresponding 2 terms for which interaction is suspected, such as $X_1 * X_2$.

PROBLEM 18.49 (PROBLEM 18.46 revisited)

This problem pertains to the amount of money withdrawn from ATM machines as it relates to the median prices of the nearby neighborhood. There is a dummy column containing a 1 if the ATM is located at a shopping center, and a 0 if it is not.

- a. If the data are not current, open the file by using the menu:
  Select **File>Other Files>Import ASCII Data**; type **c1-c3** for *column* (or whatever columns you choose) and click *OK*. Give the file name *atm2.dat* and its location. Type the heading **Cash** for c1, **Home** for c2, and **Locat** for c3.

- b. We now wish to include an interaction term for the possible relationship between *Home Median Cost* and *ATM Location*. Type the heading **Inter** for c4. To construct this new column, type
  ```
  MTB > let 'Inter' = 'Home' * 'Locat'
  ```

- c. Select **Stat>Regression>Regression...**

- d. For *Response*: double-click on *Cash;* for *Predictors* double-click on *Home* and double-click on *Locat* and double-click on *Inter*.

- e. Under *Storage* click on *Residuals*, and on *Standard.resids* and on *Fits* and on *Deleted t resids* and on *HI(leverage)* and on *Cook's Distance*. (This places some results in columns c4-c9 which could be used later.)

- f. Click on the *Options* button. Click on *Durbin-Watson statistic* and click on *Variance inflation factors* and make sure the *Fit intercept* box is checked. Click on *OK*.

- g. Click on *OK*.

- h. The entire display is shown below.

```
Regression Analysis
The regression equation is
Cash = 29.9 + 0.392 Home + 0.95 Locat + 0.0015 Inter
```

Predictor	Coef	Stdev	t-ratio	p	VIF
Constant	29.860	2.327	12.83	0.000	
Home	0.39211	0.01317	29.77	0.000	3.3
Locat	0.947	3.007	0.31	0.759	31.0
Inter	0.00154	0.01620	0.10	0.926	38.3

```
s = 1.043      R-sq = 99.6%      R-sq(adj) = 99.5%
```

Analysis of Variance

SOURCE	DF	SS	MS	F	p
Regression	3	3278.4	1092.8	1004.64	0.000
Error	11	12.0	1.1		
Total	14	3290.4			

SOURCE	DF	SEQ SS
Home	1	3273.4
Locat	1	5.1
Inter	1	0.0

Unusual Observations

Obs.	Home	Cash	Fit	Stdev.Fit	Residual	St.Resid
2	170	99.000	96.518	0.398	2.482	2.57R

R denotes an obs. with a large st. resid.

Durbin-Watson statistic = 1.95

i.   If you typed rather than used the menus, the commands follow. Note that Minitab provides headings for the standardized residuals (*SRES1* in c5), the fitted values (*FITS1* in c6), the residuals (*RESI1* in c7), the Studentized deleted residuals (*TRES1* in c8), the hat matrix elements (*HI1* in c9), and Cook's distance statistic (*COOK1* in c10). Notice that Minitab typed the headings for some new columns.

```
MTB > Name c5 = 'SRES1' c6 = 'FITS1' c7 = 'RESI1' c8 = 'TRES1' c9 =
      'HI1' &
CONT>       c10 = 'COOK1'
MTB > Regress 'Cash' 3 'Home' 'Locat' 'Inter';
SUBC>    SResiduals 'SRES1';
SUBC>    Fits 'FITS1';
SUBC>    Constant;
SUBC>    Residuals 'RESI1';
SUBC>    Tresiduals 'TRES1';
SUBC>    Hi 'HI1';
SUBC>    Cookd 'COOK1';
SUBC>    VIF.
SUBC>    DW.
```

j.   The regression equation is shown to be
Cash = 29.9 + 0.392 Home + 0.95 Locat + 0.0015 Inter .
The interpretation of this result is not as straight-forward as when there is no interaction cross-product term: in this case not only does the value of *Home* affect *Cash*, and not only does the value of *Locat* affect *Cash*, but the way in which *Home* and *Locat* interact with each other also affects *Cash*.

Multicollinearity
If the independent variables are themselves correlated (in addition to being correlated with the dependent variable, as is expected), this can cause problems in the analysis and the validity of the data.

PROBLEM 18.54
This problem pertains to the first problem of this chapter, problem 18.3 .
There are 24 months of data, where the Distribution Cost is assumed to be dependent on Sales and Number of orders.

a.   Part of the display is shown below.
```
Regression Analysis
The regression equation is
Cost = - 2.73 + 0.0471 Sales + 0.0119 Orders

Predictor      Coef      Stdev    t-ratio        p       VIF
Constant     -2.728      6.158      -0.44    0.662
Sales       0.04711    0.02033       2.32    0.031       2.8
Orders     0.011947   0.002249       5.31    0.000       2.8

s = 4.766      R-sq = 87.6%      R-sq(adj) = 86.4%
```

b.   Note that the VIF (variance inflationary factor) is shown in the table above as 2.8 for both independent variables. With the value of 2.8, there is no reason to suspect that the independent variables are themselves correlated.

Influence Analysis in Multiple Regression

The goal is to test whether some observations have undue influence on the analysis, opening up the possibility of discovering some errors in the data, or perhaps repeating some analyses after removing some data.

PROBLEM 18.58

This problem pertains to the first problem of this chapter, problem 18.3 .

There are 24 months of data, where the Distribution Cost is assumed to be dependent on Sales and Number of orders.

a.   The entire display is shown below.

```
Regression Analysis
The regression equation is
Cost = - 2.73 + 0.0471 Sales + 0.0119 Orders

Predictor        Coef        Stdev     t-ratio         p        VIF
Constant       -2.728        6.158       -0.44     0.662
Sales          0.04711     0.02033        2.32     0.031        2.8
Orders         0.011947    0.002249       5.31     0.000        2.8

s = 4.766       R-sq = 87.6%      R-sq(adj) = 86.4%

Analysis of Variance
SOURCE          DF          SS           MS          F          p
Regression       2        3368.1       1684.0      74.13      0.000
Error           21         477.0         22.7
Total           23        3845.1

SOURCE          DF        SEQ SS
Sales            1        2726.8
Orders           1         641.3

Unusual Observations
Obs.     Sales        Cost        Fit   Stdev.Fit   Residual    St.Resid
  1        386      52.950     63.425       1.332    -10.475      -2.29R

R denotes an obs. with a large st. resid.

Durbin-Watson statistic = 2.26

Fit   Stdev.Fit       95.0% C.I.            95.0% P.I.
69.878     1.663   ( 66.419,  73.338)   ( 59.378,  80.379)
```

b.   The *hat matrix* results were obtained in the initial Minitab command. A particular *hat matrix* value is considered a candidate for a closer look if it exceeds 4/n, which for this problem is 4/24 = .1667. Months 7, 12, 14, 17, 18, 19, and 20 each has a value which exceeds .1667 (see the shaded cells in the table which follows, in the *hat matrix* column), so these months might be removed and the analysis run again.

c.   The *Studentized Deleted Residuals* were also obtained in the initial Minitab command. A *Studentized Deleted Residual* value is considered a candidate for a closer look if its absolute value exceeds $t_{.95,n-p-1}$, and n-3-1 for this case is 20. Type
```
MTB > invcdf .95 ;
SUBC> t 20 df .
```

d.   The display is shown below, so the t value of interest is $t_{.95,n-3-1} = 1.7247$. Note in the table below (in the *Studentized Deleted Residual* column) that months 1, 2, and 14 have a Studentized Deleted Residual whose absolute value exceeds 1.7247, and these values are shaded in the table. Therefore, these months might be removed and the analysis run again.
```
Inverse Cumulative Distribution Function
Student's t distribution with 11 d.f.
  P( X <= x)              x
  0.9500                1.7247
```

e.   The *Cook's Distance Statistic* values were also obtained in the initial Minitab command. A *Cook's Distance Statistic* value is considered a candidate for a closer look if it exceeds $F_{.50,p,n-p}$, which for this example is $F_{.50,3,24-3}$. Therefore, type
```
MTB > invcdf .50 ;
SUBC> F 3 21 df .
```

f.   The display is shown below, so the F value of interest is $F_{.50,3,24-3} = 0.8149$. Looking at the table below (in the *Cook's Distance* column), no values exceed 0.8149, so no value has an undue influence in terms of the *Cook's Distance* value.
```
Inverse Cumulative Distribution Function
F distribution with 3 d.f. in numerator and 21 d.f. in
   denominator
P( X <= x)              x
   0.5000             0.8149
```

g.   The data are shown below. This table was generated by Minitab at the initial command. The shaded cells pertain to observations which might exert undue influence, and therefore might be discarded and the analysis repeated.

	X	Y	Standardized Residuals		Residuals	Studentized Deleted Residuals	Hat Matrix	Cook's Distance
Cost	Sales	Order s	SRES1	FITS1	RESI1	TRES1	HI1	COOK1
52.95	386	4015	-2.28885	63.4246	-10.4746	-2.57832	0.078056	0.147848
71.66	446	3806	1.74888	63.7545	7.9055	1.84649	0.100514	0.113929
85.58	512	5309	0.16999	84.8203	0.7597	0.16601	0.120724	0.001322
63.69	401	4262	-0.74126	67.0822	-3.3922	-0.73305	0.078103	0.015517
72.81	457	4296	0.57576	70.1268	2.6832	0.56637	0.043922	0.005076
68.44	458	4097	0.13946	67.7965	0.6435	0.13616	0.062639	0.000433
52.46	301	3213	0.61513	49.8385	2.6215	0.60579	0.200492	0.031629
70.77	484	4809	-1.46013	77.5276	-6.7576	-1.5033	0.057098	0.043035
82.03	517	5237	-0.47959	84.1957	-2.1657	-0.47062	0.102348	0.008742
74.39	503	4732	-0.67215	77.5029	-3.1129	-0.66312	0.05581	0.008902
70.84	535	4413	-0.9914	75.1995	-4.3595	-0.99097	0.148795	0.05727
54.08	353	2921	1.26096	48.7999	5.2801	1.27998	0.22814	0.156656
62.98	372	3977	0.14769	62.311	0.669	0.14421	0.096861	0.00078
72.3	328	4428	1.74597	65.6261	6.6739	1.84289	0.356792	0.563655
58.99	408	3964	-1.05123	63.8518	-4.8618	-1.05401	0.05841	0.022851
79.38	491	4582	0.91166	75.1455	4.2345	0.90784	0.050269	0.014664
94.44	527	5582	1.30688	88.7885	5.6515	1.33064	0.176774	0.12225
59.74	444	3450	0.07828	59.4072	0.3328	0.07641	0.20447	0.000525
90.5	623	5079	0.79229	87.3021	3.1979	0.78502	0.282845	0.082525
93.24	596	5735	-0.14655	93.8672	-0.6272	-0.14309	0.193591	0.001719
69.33	463	4269	-0.1628	70.0869	-0.7569	-0.15898	0.048453	0.00045
53.71	389	3708	-1.3537	59.8983	-6.1883	-1.38279	0.080074	0.05317
89.18	547	5387	0.39806	87.4011	1.7789	0.38994	0.120852	0.00726
66.8	415	4161	0.05713	66.5352	0.2648	0.05576	0.053968	0.000062

Stepwise Regression

Minitab provides a Stepwise regression method, which tries to find the best model (the best selection of variables) without testing all possible combinations. The menu approach is **Stat>Regression>Stepwise**. The typical approach is to add and remove variables based on the F-statistic which is calculated for each variable in the model.

Best Subset Regression

Minitab also provides a Best Subset method, which looks for the best model with, say, one independent variable, and the best model with two independent variables, and so on. The menu approach is **Stat>Regression>Best Subsets**. The typical approach is to add and remove variables based on r^2 or an adjusted r^2.

TIME-SERIES FORECASTING

This chapter introduces a variety of time-series models for forecasting purposes.

PROBLEM 19.2

The data are annual earnings per share for TRW company, over 23 years.

Obtain a plot

  a. Open the file by using the menu **File>Other Files>Import ASCII Data**; type **c1-c2** for *column* (or whatever columns you choose) and click *OK*. Give the file name *TRW.dat* and its location. Type the heading **Index** for c1 and **EPS** for c2.

  b. Select **Graph>Time Series Plot...**

  c. For *Graph Variables*: for *Y* double-click on *EPS*

  d. For *Time Scale* click on *Calendar* and use the arrow to select *Years*

  e. Title: Click on **Annotation>Title** and type **Plot of Earnings per Share with Year 1 = 1970**

  f. Click in the first box under *Text Size* and type **1** to reduce the title size. Click *OK*.

  g. Click on the *Options* button and under *Start time* type **70** for the initial year. Click *OK*.

  h. Click on *OK* to obtain the graph.

i.   (Note: using the graph editor I removed alternate years labels along the x-axis.)

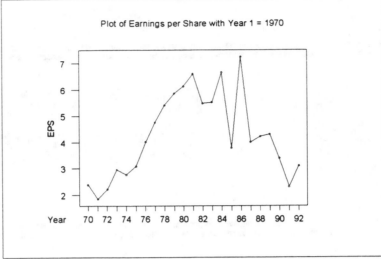

j.   If you typed rather than used the menus,

```
MTB > TSPlot 'EPS';
SUBC>    Year;
SUBC>    TDisplay 11;
SUBC>    Start 70;
SUBC>    Symbol;
SUBC>    Connect;
SUBC>   Title "Plot of Earnings per Share with Year 1 = 1970";
SUBC>     TSize 1;
SUBC>    Tick 11;
SUBC>    Tick 2.
```

3-Year Moving Average

a.   Select **Stat>Time Series>Moving Average...**

b.   For *Variable*: double-click on *EPS*

c.   For *MA Length:* type **3**

d.   Click on the box *Center the moving averages*

e.   Title: type **3-Year Moving Average: Year 1 = 1970**

f.   (Note that the following is optional: this will cause the results to be placed in the worksheet.) Click on the *Storage* button: click on *Moving averages,* click on *Fits,* and click on *Residuals.* Click *OK.*

g.   Click on the *Options* button and click on *Plot smoothed vs. actual* and click on *Summary table and results table.* Click *OK.*

(Note that including the *results table* writes the results to the Session Window; if this is not desired, don't check this box.)

h.   Click on *OK* to obtain the graph.

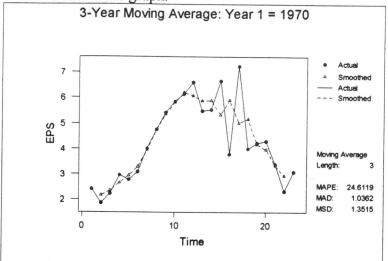

i.   If you typed rather than used the menus,

```
MTB > Name c6 = 'AVER2' c7 = 'FITS2' c8 = 'RESI2'
MTB > %MA 'EPS' 3;
SUBC>   Center;
SUBC>     Title '3-Year Moving Average: Year 1 = 1970';
SUBC>     Averages 'AVER2';
SUBC>     Fits 'FITS2';
SUBC>     Residuals 'RESI2';
SUBC>     Smplot;
SUBC>     Table.
```

Exponentially smoothed average, with the smoothing constant of $\alpha = .50$

a.   Select **Stat>Time Series>Single Exp Smoothing...**

b.   For *Variable*: double-click on *EPS*

c.   For *Weight to use in smoothing* click on *Use* and in the box type **.50**

d.   Click on the *Generate forecasts* box, and for *number of forecasts* type **1** to get the forecast for 1993

e.   Title: type **.50 Exponential Smoothing, Year 1 = 1970**

f.   (Note that the following is optional: this will cause the results to be placed in the worksheet.) Click on the *Storage* button: click on *Smoothed data,* click on *Fits,* and click on *Residuals.* Click *OK.*

g.   Click on the *Options* button and click on *Plot smoothed vs. actual*
     (Optional: click on *Summary table and results table.* Including the *results table* writes the results to the Session Window; if this is not desired, don't check this box.)

h.   For *Set initial smoothed value, use average of first,* type **1** so the first value is also the first smoothed value. Click *OK.*

i.   Click on *OK* to obtain the graph.

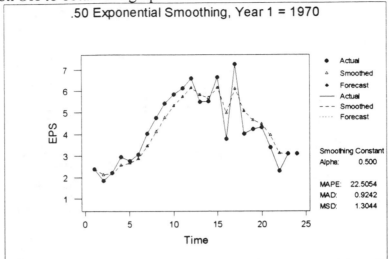

j.   The display is shown below. (Some of the results are also printed on the graph.)

```
Single Exponential Smoothing
Data      EPS
Length    23.0000
NMissing  0

Smoothing Constant
Alpha: 0.5

Accuracy Measures
MAPE: 22.5054
MAD:   0.9242
MSD:   1.3044

 Row  Period  Forecast      Lower      Upper
  1     24    3.10351    0.839108    5.36792
```

k.   If you typed rather than used the menus,
```
MTB > Name c4 = 'SMOO1' c5 = 'FITS1' c6 = 'RESI1'
MTB > %SES 'EPS';
SUBC>    Weight .50;
SUBC>    Forecasts 1;
SUBC>    Title '.50 Exponential Smoothing, Year 1 = 1970';
SUBC>    Smoothed 'SMOO1';
SUBC>    Fits 'FITS1';
SUBC>    Residuals 'RESI1';
SUBC>    Smplot;
SUBC>    Initial 1.
```

l.   The forecasted smoothed value for 1993, the next period, is shown in the output.
      The smoothed forecast for period 24 (1993) is 3.10351 .

Time-Series Analysis of Annual Data: Least Squares Trend Fitting and Forecasting

PROBLEM 19.11

The data are the receipts and expenditures for state and local governments over a 22-year period. We'll only be interested in the Receipts portion of the data.

a.   Open the file by using the menu **File>Other Files>Import ASCII Data**; type **c1-c4** for *column* (or whatever columns you choose) and click *OK*. Give the file name *state.dat* and its location. Type the heading **Index** for c1, **Receipt** for c2, **Expend** for c3, and **Diff** for c4.

We want to determine the Trend for the *Receipts.*

a.   Select **Stat>Time Series>Trend Analysis...**
b.   For *Variable*: double-click on *Receipt*
c.   For *Model Type* click on *Linear*
d.   Click on the *Generate forecasts* box, and for *number of forecasts* type **3** to get the forecast for 1992-1994
e.   Title: type **Trend for Receipts, Year 1 = 1970**
f.   (Note that the following is optional: this will cause the results to be placed in the worksheet.) Click on the *Storage* button: click on *Fits*, and click on *Residuals*. Click *OK*.
g.   Click on the *Options* button and click on *Display plot*
h.   Click on *OK* to obtain the graph.

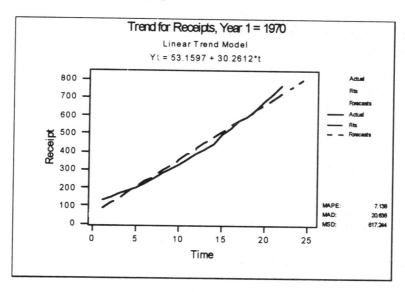

i. The display is shown below. The linear equation is $Y_t = 53.1597 + 30.2612*t$, where t is the index for which month. (Important point: Minitab assumes that the first year has an index of 1, and the second year has an index of 2, and so on. If instead the index for years begins with 0, then the constant would differ, and the equation would be $Y_t = 83.42 + 30.2612*t$. We'll proceed with Minitab's assumption that the index values begin at 1 rather than 0.)

```
Trend Analysis
Data      Receipt
Length    22.0000
NMissing  0

Fitted Trend Equation
Yt = 53.1597 + 30.2612*t

Accuracy Measures

MAPE:        7.13835
MAD:         20.6355
MSD:         617.244

  Row  Period  Forecast
   1      23    749.167
   2      24    779.429
   3      25    809.690
```

j. The forecasted values are shown above: for 1992: 749.167; for 1993: 779.429; and for 1994: 809.690.

k. If you typed rather than used the menus,
```
MTB > Name c5 = 'FITS1' c6 = 'RESI1'
MTB > %Trend 'Receipt';
SUBC>    Forecasts 3;
SUBC>    Title 'Trend for Receipts, Year 1 = 1970';
SUBC>    Fits 'FITS1';
SUBC>    Residuals 'RESI1'.
```

Quadratic Trend Equation for the *Receipts*

We want to determine the Quadratic Trend Equation for the *Receipts*. This uses exactly the same instructions as before, except for *Model Type* click on *Quadratic* and changing the title. However, all the instructions follow.

a. Select **Stat>Time Series>Trend Analysis...**
b. For *Variable*: double-click on *Receipt*
c. For *Model Type* click on *Quadratic*
d. Click on the *Generate forecasts* box, and for *number of forecasts* type **3** to get the forecast for 1992-1994
e. Title: type **Quadratic for Receipts, Year 1 = 1970**

f.   (Note that the following is optional: this will cause the results to be placed in the worksheet.)   Click on the *Storage* button: click on *Fits*, and click on *Residuals*. Click *OK*.

g.   Click on the *Options* button and click on *Display plot*
(Optional: click on *Summary table and results table*. Including the *results table* writes the results to the Session Window; if this is not desired, don't check this box.)

h.   Click on *OK* to obtain the graph.

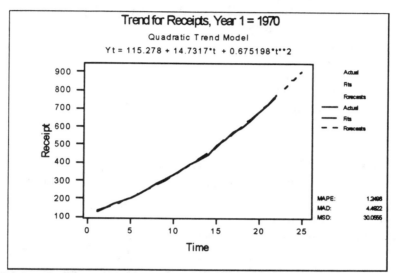

i.   The display is shown below.   The Quadratic Trend Equation is Y_t = 115.278 + 14.7317*t + 0.675798*t^2, where t is the index for which month.   (Important point: Minitab assumes that the first year has an index of 1, and the second year has an index of 2, and so on. If the index for years begins with 0, then the constant would differ, and the equation would be Y_t = 130.865 + 16.082*t + 0.6752*t^2. Interestingly, the forecasts are the same regardless of which equation is used.   We'll proceed with Minitab's assumption that the index values begin at 1 rather than 0.)

```
Trend Analysis
Data       Receipt
Length     22.0000
NMissing 0

Fitted Trend Equation
Yt = 115.278 + 14.7317*t  + 0.675198*t**2

Accuracy Measures
MAPE:          1.24957
MAD:           4.48217
MSD:          30.0555
```

```
Row   Period   Forecast
 1      23     811.286
 2      24     857.752
 3      25     905.568
```

j.   The forecasted values are shown above: for 1992: 811.286; for 1993: 857.752; and for 1994: 905.568.

k.   If you typed rather than used the menus,
```
MTB > Name c9 = 'FITS2' c10 = 'RESI2'
MTB > %Trend 'Receipt';
SUBC>    Quadratic;
SUBC>    Forecasts 3;
SUBC>    Title 'Trend for Receipts, Year 1 = 1970';
SUBC>    Fits 'FITS2';
SUBC>    Residuals 'RESI2'.
```

The Holt-Winters Method for Trend Fitting and Forecasting.

This method adds a trend component to the exponential smoothing method previously used.

PROBLEM 19.19 (revisiting Problem 19.2)

The data are annual earnings per share for TRW company, over 23 years.

a.   Open the file by using the menu **File>Other Files>Import ASCII Data**; type **c1-c2** for *column* (or whatever columns you choose) and click *OK*. Give the file name *TRW.dat* and its location. Type the heading **Index** for c1 and **EPS** for c2.

b.   We want to use the Holt-Winter's method with U=.30 to smooth the level or data, and V=.30 to smooth the trend.

c.   Select **Stat>Time Series>Double Exp Smoothing...**

d.   For *Variable*: double-click on *EPS*

e.   For *Weights to use in smoothing* click on *Use* and type **.30** for *level* and type **.30** for *trend*

f.   Click on the *Generate forecasts* box, and for *number of forecasts* type **4** to get the forecasts for 1993-1996

g.   Title: type **Holt-Winter's for EPS, Year 1 = 1970**

h   (You could click on the *Storage* button and store results in the worksheet; we will ignore this.)

i.   Click on the *Options* button and click on *Plot smoothed vs. actual* (Optional: click on *Summary table and results table*. Including the *results table* writes the results to the Session Window; if this is not desired, don't check this box.)

j.   Click on *OK* to obtain the graph.

k. The forecasted values for 1993-1996 have 95% confidence bands shown as dotted lines.

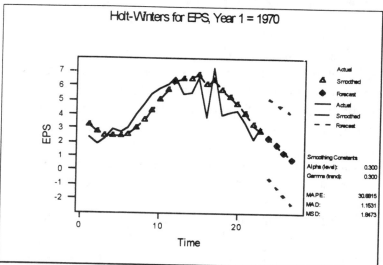

l. The display is shown below.
    Double Exponential Smoothing
    Data   EPS
    Length   23.0000
    NMissing 0

    Smoothing Constants
    Alpha (level): 0.3
    Gamma (trend): 0.3

    Accuracy Measures
    MAPE: 30.8815
    MAD: 1.1531
    MSD: 1.8473

Row	Period	Forecast	Lower	Upper
1	24	2.35371	-0.47144	5.17887
2	25	1.84371	-1.11569	4.80312
3	26	1.33372	-1.77291	4.44035
4	27	0.82372	-2.44135	4.08879

m. The forecasted values are shown above: for 1993: 2.35371; for 1994: 1.84371; for 1995: 1.33372; and for 1996: 0.82372 .

m. If you typed rather than used the menus,
    MTB > %DES 'EPS';
    SUBC>   Weight 0.3 0.3;
    SUBC>   Forecasts 4;
    SUBC>   Title 'Holt-Winters for EPS, Year 1 = 1970';
    SUBC>   Smplot.

Autoregressive Modeling for Trend Fitting and Forecasting

PROBLEM 19.30 (revisiting Problem 19.2)
The data are annual earnings per share for TRW company, over 23 years.

a. Open the file by using the menu **File>Other Files>Import ASCII Data**; type **c1-c2** for *column* (or whatever columns you choose) and click *OK*. Give the file name *TRW.dat* and its location. Type the heading **Index** for c1 and **EPS** for c2.

b. We want to fit a third-order autoregressive model and test for the significance of the parameter. If necessary, repeat for second-order and first-order models.

c. Note that Minitab has an autoregressive method (**Stat>Time Series>ARIMA...**), but here we'll use the usual regression method to obtain a third-order and then a second-order model.

d. In order to perform regression on these data, new columns have to be generated: *LAG1* for instance takes the *EPS* data and moves it down one (and deletes the last one so the column length is the same as for the original column). *LAG2* takes the original *EPS* data and moves it down two (and deletes the last two), and similarly for *LAG3*.

e. Provide headings for new columns: type **LAG1** for c3, **LAG2** for c4, and **LAG3** for c5.

f. Select **Stat>Time Series>Lag**, to store the data of *EPS* in *LAG1*, but shifted down one row:

g. For *Series* double-click on *EPS*

h. For *Lags in* double-click on *LAG1*

i. For *Lag* type **1**. Click *OK*.

j. Add the new column for *LAG2*: Select **Stat>Time Series>Lag**, For *Lags in* double-click on *LAG2*, For *Lag* type **2**. Click *OK*.

k. Add the new column for *LAG3*: Select **Stat>Time Series>Lag**, For *Lags in* double-click on *LAG3*, For *Lag* type **3**. Click *OK*.

l. Now perform the regression: Select **Stat>Regression>Regression**; for *Response* double-click *EPS*; for *Predictors* double-click *LAG1* then double-click *LAG2* then double-click *LAG3*. Click *OK*.

m. If you typed rather than used the menus
```
MTB > Regress 'EPS' 3  'LAG1' 'LAG2' 'LAG3';
SUBC>    Constant.
```

n. The display is shown below.

```
Regression Analysis
The regression equation is
EPS = 1.34 + 0.234 LAG1 + 0.471 LAG2 + 0.012 LAG3

20 cases used 3 cases contain missing values
```

Predictor	Coef	Stdev	t-ratio	p
Constant	1.3395	0.9365	1.43	0.172
LAG1	0.2344	0.2483	0.94	0.359
LAG2	0.4710	0.2251	2.09	0.053
LAG3	0.0116	0.2417	0.05	0.962

s = 1.167   R-sq = 46.4%   R-sq(adj) = 36.3%

Analysis of Variance

SOURCE	DF	SS	MS	F	p
Regression	3	18.847	6.282	4.62	0.016
Error	16	21.776	1.361		
Total	19	40.623			

SOURCE	DF	SEQ SS
LAG1	1	11.857
LAG2	1	6.987
LAG3	1	0.003

o   The third-order model equation is therefore:
$$\hat{Y}_i = 1.3395 + 0.2344\hat{Y}_{i-1} + 0.4710\hat{Y}_{i-2} + 0.0116\hat{Y}_{i-3}.$$

p.   Is the third-order parameter significant? Its t-ratio or t_{sample} = 0.05, and its p-value is .962, which is very large. Therefore, this term is not significant, and it can be dropped from the model.

q.   Now for the second-term model: perform the regression with the third term removed, so LAG3 will not be used. Select **Stat>Regression>Regression;** for *Response* double-click *EPS*; for *Predictors* double-click *LAG1* then double-click *LAG2*. (Or, just remove *LAG3* from the list.) *Click OK.*

r   If you typed rather than used the menus
```
MTB > Regress 'EPS' 3  'LAG1' 'LAG2';
SUBC>   Constant.
```

s.   The display is shown below.
```
Regression Analysis
The regression equation is
EPS = 1.14 + 0.270 LAG1 + 0.485 LAG2

21 cases used 2 cases contain missing values
```

Predictor	Coef	Stdev	t-ratio	p
Constant	1.1384	0.8031	1.42	0.173
LAG1	0.2697	0.1980	1.36	0.190
LAG2	0.4851	0.1987	2.44	0.025

s = 1.110   R-sq = 51.7%   R-sq(adj) = 46.3%

```
Analysis of Variance
SOURCE       DF          SS          MS          F          p
Regression    2      23.755      11.877      9.64      0.001
Error        18      22.186       1.233
Total        20      45.941

SOURCE       DF      SEQ SS
LAG1          1      16.407
LAG2          1       7.348
```

t. The second-order model equation is therefore:
$$\hat{Y}_i = 1.1384 + 0.2697\hat{Y}_{i-1} + 0.4851\hat{Y}_{i-2}.$$
Is the second-order parameter significant? Its t-ratio or $t_{sample} = 2.44$, and its p-value is 0.025, which is very small. Therefore, this term is significant, and it needs to remain in the model.

u. Shown below is the plot of the original data and the second-order autoregressive model. This was produced by generating a new column for the estimates: call it *2ndOrd*, and it is populated by typing at the Session Window at the MTB > prompt:
```
let '2ndOrd' = 1.1384 + .2697*'LAG1' + .4851*'LAG2'
```

v. Then the plot (**Graph>Plot**) is produced so that there are two plots: *EPS* vs *Index* and *2ndOrd* vs *Index*. The two plots must overlap so click on *Frame, Multiple Graphs, Overlap graphs on the same page*.

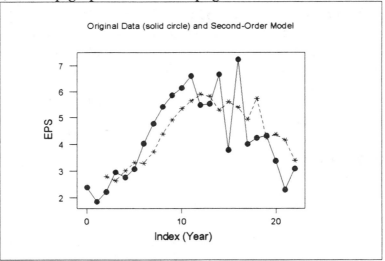

w. If you typed rather than used the menus,
```
MTB > Plot 'EPS'*'Index' '2ndOrd'*'Index';
SUBC>    Symbol;
SUBC>      Type 6 4;
SUBC>    Connect;
SUBC>    Title "Original Data (solid circle) and Second-Order Model";
SUBC>      TSize 1;
SUBC>    Overlay.
```

Time-Series Forecasting of Monthly Data

PROBLEM 19.46

The data represent the monthly outlays (in thousands of dollars) to a municipality's sanitation department.

- a. Open the file by using the menu **File>Other Files>Import ASCII Data**; type **c1-c2** for *column* (or whatever columns you choose) and click *OK*. Give the file name *outlays.dat* and its location. Type the heading **Index** for c1 and **Outlay** for c2.
- b. Select **Stat>Time Series>Trend Analysis...**
- c. For *Variable*: double-click on *Outlay*
- d. For *Model Type* click on *Linear*
- e. Title: type **Trend for Outlay, Year 1 = 1985**
- f. (Note that the following is optional: this will cause the results to be placed in the worksheet.) Click on the *Storage* button: click on *Fits*, and click on *Residuals*. Click *OK*.
- g. Click on the *Options* button and click on *Display plot*
- h. Click on *OK* to obtain the graph.

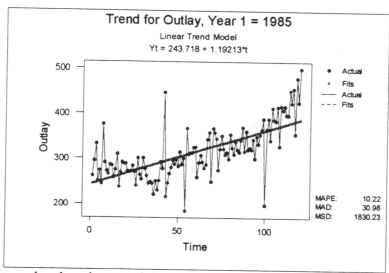

- i. If you typed rather than used the menus,
  ```
  MTB > %Trend 'Outlay';
  SUBC>    Title 'Trend for Outlay, Year 1 = 1985'.
  ```

j.   The display is shown below. The Trend equation is $Y_t = 243.718 + 1.19213*t$

```
Trend Analysis
Data       Outlay
Length     120.000
NMissing 0


Fitted Trend Equation

Yt = 243.718 + 1.19213*t

Accuracy Measures

MAPE:         10.2229
MAD:          30.9779
MSD:          1830.23
```

Seasonal Index

Next we wish to determine the seasonal index. This can be obtained directly from Minitab.

a.   Select **Stat>Time Series>Decomposition...**
b.   For *Variable*: double-click on *Outlay*
c.   For *Model Type* click on *Multiplicative*
d.   For *Seasonal length* type **12**
e.   For *Model components* click on *Trend plus seasonal*
f.   Make sure *Initial seasonal period* has the value 1.
g.   Click on *Generate Forecasts* and for *Number of forecasts* type **12**
h.   Title: type **Seasonal Decomposition for Outlay**
i.   Click on *OK* to obtain the graph.

j.   If you typed rather than used the menus

```
MTB > %Decomp 'Outlay' 12;
SUBC>   Forecasts 12;
SUBC>   Title 'Seasonal Decomposition for Outlay';
SUBC>   Start 1.
```

k.   The display is shown below. The Trend equation is $Y_t = 243.718 + 1.19213*t$, as
was determined earlier when we did the *Trend* analysis.

```
Time Series Decomposition
Data       Outlay
Length     120.000
NMissing 0

Trend Line Equation
Yt = 243.718 + 1.19213*t
```

```
Seasonal Indices
  Period     Index
     1      0.940707
     2      0.984687
     3      1.07180
     4      0.964167
     5      0.964037
     6      1.04502
     7      1.03255
     8      1.02756
     9      0.985155
    10      0.963867
    11      1.02405
    12      0.996399
```

Accuracy of Model

```
MAPE:        9.91
MAD:        29.78
MSD:      1715.19
```

Forecasts

```
  Row    Period    Forecast
    1     121      364.962
    2     122      383.198
    3     123      418.378
    4     124      377.512
    5     125      378.610
    6     126      411.662
    7     127      407.977
    8     128      407.234
    9     129      391.601
   10     130      384.289
   11     131      409.502
   12     132      399.634
```

l.   The table below provides the forecast for the next year, 1995. The *TREND* forecast was obtained when the Trend analysis was completed. The Seasonal Index was obtained when the *Decomposition* method was used. The *Adjusted Forecast* was also obtained from the *Decomposition* method, although it could have been computed separately, since the *Adjusted Forecast* column is just the *Trend Forecast* column multiplied by the *Seasonal Index*.

Month	Period	TREND Forecast	Seasonal Index	Adjusted Forecast
January	121	387.965	0.940707	364.962
February	122	389.158	0.984687	383.198
March	123	390.350	1.071800	418.378
April	124	391.542	0.964167	377.512
May	125	392.734	0.964037	378.610
June	126	393.926	1.045020	411.662
July	127	395.118	1.032550	407.977
August	128	396.310	1.027560	407.234
September	129	397.502	0.985155	391.601
October	130	398.695	0.963867	384.289
November	131	399.887	1.024050	409.502
December	132	401.079	0.996399	399.634

m.   The *Decomposition* method of *Time-Series Analysis* when performed by Minitab automatically generates the following three graphs, which provide useful views of the process.

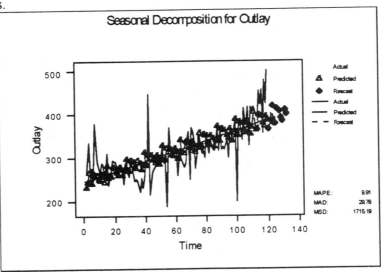

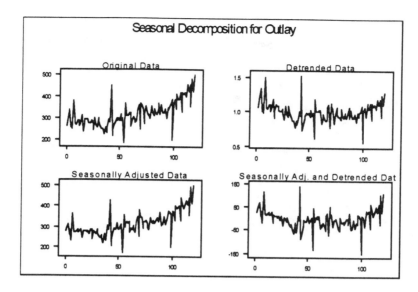

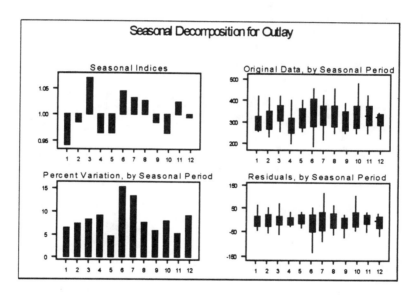